WRITING PROPOSALS AND GRANTS

WRITING PROPOSALS AND GRANTS

THIRD EDITION

Richard Johnson-Sheehan
Paul Thompson Hunter

Parlor Press
Anderson, South Carolina
www.parlorpress.com

Parlor Press LLC, Anderson, South Carolina, USA
© 2024 by Parlor Press.
All rights reserved.
Printed in the United States of America on acid-free paper.

S A N: 2 5 4 - 8 8 7 9

Library of Congress Cataloging-in-Publication Data on File

1 2 3 4 5

978-1-64317-475-4 (paperback)
978-1-64317-476-1 (PDF)
978-1-64317-477-8 (EPUB)

Cover and interior design by David Blakesley.
Photo by Google DeepMind on Unsplash. An artist's illustration of artificial intelligence (AI). This image explores how AI-generated media can be watermarked. It was created by Zünc Studio as part of the Visualising AI project launched by Google DeepMind.

Parlor Press, LLC is an independent publisher of scholarly and trade titles in print and multimedia formats. This book is available in print and ebook formats from Parlor Press on the World Wide Web at parlorpress.com or through online and brick-and-mortar bookstores. For submission information or to find out about Parlor Press publications, write to Parlor Press, 3015 Brackenberry Drive, Anderson, South Carolina, 29621, or email editor@parlorpress.com.

CONTENTS

Preface — xii
  Premises of This Book — xii
  Content and Organization — xiii
  New Material in the Third Edition — xiv
  Acknowledgments — xiv

1 An Introduction to Proposals and Grants — 3
  Why Write Proposals? — 3
  The Power of Persuasion — 4
    Interpreting an Evolving Reality — 4
    Expressing Your Ideas Persuasively — 4
  What Is a Proposal Anyway? — 5
    A Proposal Is a Genre — 5
    Categories of Proposals — 7
  The Proposal-Writing Process — 7
  Where Does Artificial Intelligence Fit In? — 8
  Grant Proposals — 9
  Chapter Summary and Looking Ahead — 10
  Try This Out! — 10
  Case Study: The Carbon Neutral Campus Project—
  Introductions — 11

2 Requests for Proposals (RFPs) — 13
  Two Basic Reasons for Writing Proposals — 13
    Reading RFPs — 14
      Two Example RFPs — 14
      Interpreting an RFP — 17
  Determining the Stasis of an Opportunity — 17
    Using the Five-W and How Questions — 17
    Using Proposal Opportunity Worksheets — 18
  Defining the Problem or Opportunity — 20
    Step 1: Is There a Problem? — 20
    Step 2: What Exactly Is the Problem? — 22
    Step 3: How Serious Is the Problem? — 23
    Step 4: What Kind of Proposal Is Needed to Solve the Problem? — 23
      Research Proposals — 23
      Planning Proposals — 24
      Implementation Proposals — 24
      Estimate Proposals (Sales Proposals) — 24
  Talking to the Point of Contact — 25

Chapter Summary and Looking Ahead	27
Try This Out!	27
Case Study: The Carbon Neutral Campus Project— What Is the Problem?	29

3 Getting Started on Your Proposal — 33

Topic, Angle, Purpose, Readers, and Context	33
Topic: What Is Need to Know vs. Want to Tell Information?	35
Angle: What Is New or Has Changed Recently?	35
Purpose: What Is the Project's Primary Objective?	36
Reader Analysis: Who Can Say Yes to the Proposal?	36
Primary Readers (Decision Makers)	37
Secondary Readers (Advisors)	37
Tertiary Readers (Evaluators)	37
Gatekeepers (Supervisors)	37
Motives	39
Values	40
Attitudes	40
Emotions	40
Context Analysis: What External Factors Will Influence the Readers?	42
Physical Context	42
Economic Context	42
Ethical Context	43
Political Context	43
Focusing a Writing Team	45
Chapter Summary and Looking Ahead	45
Try This Out!	46
Case Study: The Carbon Neutral Campus Project— What Is the Rhetorical Situation?	46

4 Writing the Introduction — 52

Telling Them What You're Going to Tell Them	52
Move 1: Grab Your Readers' Attention	53
Move 2: Define the Topic of Your Proposal	53
Move 3: Provide Background Information about the Topic	54
Move 4: Stress the Importance of Your Proposal's Topic to the Readers	54
Move 5: State the Purpose (Primary Objective) of Your Proposal	55
Move 6: State Your Proposal's Main Point	55
Move 7: Forecast the Organization of the Proposal	55
Working Out the Seven Moves on a Worksheet	56
Drafting Your Proposal's Introduction	56
Putting the Moves in Order	56
Revising the Introduction to Make it Clearer and More Concise	58
Chapter Summary and Looking Ahead	58
Try This Out!	59

Case Study: The Carbon Neutral Campus Project—
How Do We Get Started? — 59

5 Writing the Background section — 62

- Describing the Current Situation — 62
- Guidelines for Drafting the Background Section — 63
  - Guideline 1: Problems Are the Effects of Causes — 63
  - Guideline 2: Ignored Problems Tend to Grow Worse — 66
  - Guideline 3: Blame Change, Not People — 67
  - Mapping in Teams — 68
- Researching the Current Situation — 68
- Using Artificial Intelligence to Generate Content — 69
- Drafting the Background Section — 70
  - Opening — 70
  - Body Paragraphs — 70
    - Causal Approach — 70
    - Effects Approach — 71
    - Narrative Approach — 71
  - Closing — 72
- Special Case: Research Grants and Literature Reviews — 72
  - Literature Review — 73
  - Prior Research — 73
- Chapter Summary and Looking Ahead — 74
- Try This Out! — 75
- Case Study: The Carbon Neutral Campus Project—
What is the Problem? — 75

6 Writing the Project Plan or Methods Section — 80

- The Project Plan or Methods Section — 80
- Setting Objectives for the Project Plan — 80
  - Using the Objectives Provided by the Customer, Client, or Funding Source — 82
  - Developing Your Own Objectives — 82
- Answering the *How* Questions — 83
  - Step 1: Write Down A Few Solutions to the Problem — 83
  - Step 2: Map the Major Steps of a Possible Solution — 83
  - Step 3: Map the Minor Steps for Each Major Step — 84
  - Step 4: Determine if Your Plan Can Achieve the Project's Objectives — 86
  - Step 5: Answer the Readers' *Why* Questions — 87
  - Step 6: State Deliverables for Each Major Step — 87
- Drafting the Project Plan Section — 88
  - The Opening of the Project Plan Section — 89
  - The Body of the Project Plan Section — 90
  - The Closing of the Project Plan Section — 91
- Developing a Project Timeline — 91
- A Comment on Research Methods Sections — 92

Chapter Summary and Looking Ahead 93
　　　Try This Out! 93
　　　Case Study: The Carbon Neutral Campus Project—
　　　What's the Project Plan? 94

7 Writing the Qualifications Section 98
　　　People Hire People, Not Proposals 98
　　　Types of Qualifications Sections 98
　　　What Makes You Different Makes You Attractive 99
　　　Drafting the Qualifications Section 102
　　　　Opening 102
　　　　Body 102
　　　　　Description of Personnel 102
　　　　　Description of the Company/Organization 104
　　　　　Experience of the Company/Organization 105
　　　　Closing 105
　　　Creating a Persona 105
　　　To Boilerplate or Not to Boilerplate 107
　　　Chapter Summary and Looking Ahead 108
　　　Try This Out! 108
　　　Case Study: The Carbon Neutral Campus—Who Can Do the Project? 109

8 Writing a Conclusion with Costs and Benefits 112
　　　Finishing on a High Note 112
　　　Identifying the Benefits 112
　　　　　Hard Benefits: Deliverables 113
　　　　　Soft Benefits: Qualities 113
　　　　　Value Benefits: Shared Worldview 114
　　　Drafting the Conclusion 115
　　　　Move 1: Make an Obvious Transition from the Body of the Proposal 116
　　　　Move 2: State the Costs of the Plan 116
　　　　　Presenting the Costs 116
　　　　　Offering a Synopsis of the Costs 117
　　　　　Stating the Costs in a Straightforward Way 118
　　　　Move 3: Summarize the Benefits of the Plan 118
　　　　Move 4: Look to the Future 119
　　　　Move 5: Thank Them and Identify the Next Step 119
　　　Chapter Summary and Looking Ahead 119
　　　Try This Out! 120
　　　Case Study: The Carbon Neutral Campus Project—
　　　What Are the Costs and Benefits? 120

9 Budgeting a Project 123
　　　Budgets: The Bottom Line Is the Bottom Line 123
　　　Budget Basics 123
　　　　Itemized and Nonitemized Budgets 123
　　　　Fixed and Flexible Budgets 124

Fixed, Variable, and Semi-Variable Costs	*125*
Developing a Budget	*126*
Management, Principal Investigators, and Salaried Labor	*126*
Direct Labor	*127*
Indirect Labor	*130*
Facilities and Equipment	*130*
Direct Materials	*131*
Indirect Materials	*131*
Travel	*131*
Communication and Marketing	*132*
Profit (For-Profit Business Proposals Only)	*133*
Facilities and Administrative Costs (F&A) (Grant Proposals Only)	*136*
Cost Sharing (Grant Proposals Only)	*136*
Matching Funds	*136*
In-Kind Contributions	*137*
Writing the Budget Rationale or Budget Section	*137*
Positioning the Budget	*137*
Organizing the Budget Rationale	*138*
Projecting Confidence	*138*
Chapter Summary and Looking Ahead	*139*
Try This Out!	*139*
Case Study: The Carbon Neutral Campus Project—How Much Will This Cost?	*140*

10 Writing with Style — *144*

Choosing to Write Clearly and Persuasively	*144*
Writing Plain Sentences	*145*
Guideline 1: The Subject Should Be What the Sentence Is About	*146*
Guideline 2: Make the "Doer" the Subject	*147*
Guideline 3: State the Action in the Verb	*147*
Guideline 4: Put the Subject Early in the Sentence	*147*
Guideline 5: Eliminate Nominalizations	*148*
Guideline 6: Avoid Excessive Prepositional Phrases	*149*
Guideline 7: Eliminate Redundancy	*149*
Guideline 8: Make Sentences Breathing Length	*149*
A Simple Method for Writing Plainer Sentences	*150*
Writing Plain Paragraphs	*151*
The Elements of a Paragraph	*151*
Transition Sentence	*151*
Topic Sentence	*151*
Support Sentences	*152*
Point Sentences	*152*
Building a Plain Paragraph	*153*
Aligning Sentence Subjects in a Paragraph	*153*
The Given/New Method	*155*
When Is It Appropriate to Use Passive Voice?	*156*
Persuasive Style	*158*

    Elevating the Tone   *158*
    Using Similes and Analogies   *161*
    Using Metaphors   *161*
        Conceptual Metaphors   *161*
        Invented Metaphors   *162*
    Changing the Pace   *162*
  Chapter Summary and Looking Ahead   *162*
  Try This Out!   *163*
  Case Study: The Carbon Neutral Campus Project—What Should Be the Proposal's Style?   *164*

11 Designing Proposals   168

  "How You Say Something . . ."   *168*
  Four Principles of Design   *168*
    Design Principle 1: Balance   *169*
        Grids that Balance a Page Layout   *171*
        Other Balance Strategies   *172*
    Design Principle 2: Alignment   *173*
    Design Principle 3: Grouping   *174*
        Headings   *174*
        Horizontal and Vertical Rules   *175*
        Borders   *175*
    Design Principle 4: Consistency   *175*
    Headers and Footers   *176*
        Typefaces   *176*
        Sequential and Nonsequential Lists   *177*
        Labeling of Graphics   *177*
  The Process of Designing Proposals   *178*
    Step 1: Consider the Rhetorical Situation   *178*
    Step 2: Thumbnail a Few Model Pages   *178*
    Step 3: Create a Design Style Sheet   *178*
    Step 4: Create a Template   *179*
    Step 5: Edit the Design   *179*
  Chapter Summary and Looking Ahead   *179*
  Try This Out!   *180*
  Case Study: The Carbon Neutral Campus Project— How Should the Proposal Be Designed?   *180*

12 Using Graphics   184

  The Need for Graphics   *184*
  Guidelines for Using Graphics   *184*
  Using Graphics to Display Information and Data   *185*
    Line Graphs   *185*
    Bar Charts   *186*
    Tables   *187*
    Pie Charts   *188*
    Figure 12.4: Pie Chart That Illustrates Percentages   *189*

  Gantt Charts   *189*
  Organizational Charts ("Org Charts")   *190*
  Photographs   *191*
  Drawings   *191*
  Other Kinds of Graphics   *192*
   A Note about AI Images   *192*
 Chapter Summary and Looking Ahead   *192*
 Try This Out!   *193*
 Case Study: The Carbon Neutral Campus Project—What Kinds of Visuals Would Help?   *193*

13 Putting Your Proposal Together   *197*
 Seeing the Proposal as a Whole Document   *197*
 Preparing the Front Matter   *197*
  Letters or Memos of Transmittal   *198*
   Writing a Letter or Memo of Transmittal   *198*
   Using Style in the Letter or Memo of Transmittal   *200*
  Cover Page   *200*
  Executive Summary   *200*
  Table of Contents   *202*
 Preparing the Back Matter   *202*
  Resumes of Management and Key Personnel   *203*
  Glossary of Terms and Symbols   *203*
  Bibliography or References List   *204*
  Formulas and Calculations   *205*
  Related Reports, Prior Proposals, and FYI Information   *205*
 Revising the Proposal or Grant   *205*
  The Rhetorical Situation   *206*
  Rethinking the Problem or Opportunity   *206*
  The Final Edit   *207*
   Content   *207*
   Organization   *207*
   Style   *208*
   Design   *208*
 Looking Ahead   *208*
 Try This Out!   *209*
 Case Study: The Carbon Neutral Campus Project—Are We Really Finished?   *209*

Appendix A: Carbon Neutral Campus Proposal   *211*

Appendix B: FusionFactory Proposal   *221*

About the Authors   *233*

Index   *235*

PREFACE

What is a proposal? In the first chapter of this book, proposals are defined as "tools for managing change." The keywords in this definition are *tools*, *managing*, and *change*. By tools, we mean proposals are devices that help people do their work. They help people present new ideas, project plans, and even their dreams for the future. Managing means taking control of a situation. It means directing people and resources in a way that allows people to achieve specific ends. Change is really what life is all about.

As a writer of proposals and grants, you should regularly remind yourself that change is always happening, offering new opportunities to take action, solve problems, and make your world better. Proposals are tools for managing these changes. They are tools for taking purposeful action in a world that never seems to stop moving.

So how do you go about using proposals and grants to manage change? The strategies in this book are rooted in rhetorical theory, a discipline going back a few thousand years. Put concisely, rhetoric is the study of what could be or what should be, not necessarily what is. It is a forward-thinking discipline, more concerned with what you are going to do in the future than what happened in the past. For this reason, the discipline of rhetoric is particularly helpful for writing successful proposals and grants. After all, rhetoric is a discipline that studies change and how humans use words to shape the evolving world around them.

Premises of This Book

A few important premises set this book apart from other books on writing proposals and grants. Other books offer mostly tips and tricks to produce more effective proposals. You will find some of that material here, too. But you will also find that this book stresses the *process* of writing successful proposals and grants. Our aim is to show you how professional writers of proposals and grants use consistent step-by-step processes to move from planning to drafting to designing to editing their work

An initial premise of this book is that writing proposals and grants should be approached from a problem-solving point of view. By seeing projects as problems to be solved, you and your team can sort out complex situations and devise plans that meet specific goals. In this book, you will learn how to use critical thinking and analysis to better understand the current situation, invent a sound plan, present your credentials, and develop a budget. You will also learn how to use style and design to make your proposals easier to read and more persuasive.

A second premise is that you, as a proposal writer or grant writer, need to pay attention to change as you assess situations and develop plans for improving those situations. Proposals are never written in a social vacuum: They are written in evolving social, political, and ethical environments that are uncertain and sometimes unpredictable. What was true today or yesterday may not be true tomorrow. Therefore, the key to writing successful proposals is to first identify what has changed to create the current problem or new opportunity. Then with these elements of change identified, you can use proposals and grants to shape the evolving situation to your advantage.

The third underlying premise of this book is that proposals should be written "visually." In this multimedia age of smart phones and computers, people tend to think in images rather than in words alone. In this book, you will notice that almost all the techniques discussed employ strategies that take advantage of your and your readers' abilities to think visually. Some of the visual techniques are more apparent, such as the use of concept maps to invent a proposal's content or the use of design principles to guide the page layout of the text. Other visual techniques are subtle, such as the use of similes and metaphors (Chapter 10) to create visual images in the readers' minds. Visual writing allows both writers and readers to develop a strong visual sense of what the proposal is illustrating.

Content and Organization

In this book, you will also learn time-tested techniques that are used by professional proposal and grant writers:

- You will learn how to tap into your innate creativity to invent the content of your proposals. Some of the "invention" techniques in this book, such as concept mapping (Chapter 5), may seem a little strange at first. However, these techniques will help you gain unique insights into problems and opportunities. They will help you devise creative strategic plans that use imagination and foresight.
- You will learn some techniques for using artificial intelligence (AI) to generate content, edit your writing, and design graphics for your document.
- You will see how to organize your ideas into familiar genres and patterns that achieve specific goals. The organizational pattern of a proposal is much more than a way to present your ideas to the readers. It is a means for helping you lay out your approach and develop a successful "capture strategy."
- You will learn how to express your ideas plainly and persuasively. Good style should be a choice, not an accident. You will learn simple, time-tested techniques for clarifying your message for the readers. You will also learn how to amplify your prose by using persuasive style techniques that tap into your readers' motives, values, attitudes, and emotions.
- You will use visual design techniques to clarify and enhance the arguments in your proposals and grants. In an increasingly visual age, readers are more reliant than ever on visual cues to help them find the important information in proposals. They also expect proposals to include graphics that illustrate and reinforce the written text. You will learn design principles that will help you devise winning designs for your proposal and include graphics to enhance the text.

This book's successive chapters will guide you through the process of composing, revising, designing, and editing proposals.

New Material in the Third Edition

The third edition of *Writing Proposals and Grants* has been revised from front to back. People who have used the second edition have been overwhelmingly positive in their feedback, and we have received excellent suggestions for sharpening some parts of the book and for adding material that readers said they would find helpful.

A major change in this edition is the discussion of using AI to accelerate and improve proposal and grant writing. In many ways, we are all learning how to use AI as a communication tool. In this book, we have done our best to explain what is possible right now and where we see the use of AI to write proposals and grants in the future.

Overall, readers who are familiar with the second edition of *Writing Proposals and Grants* should find this edition much stronger and more comprehensive in its treatment of the proposal-writing and grant-writing processes. The addition of AI will only make proposal writing and grant writing faster and more dynamic.

Acknowledgments

Let us thank the others who helped put this book together. First, we appreciate the help of our colleagues and students at Purdue University and the University of New Mexico who have shaped this text through their suggestions for improvements. At Purdue, Morgan Sousa, Allen Brizee, and Jaclyn Wells helped me strengthen the chapters on strategic planning and teaming. Allen added his expertise on stasis theory. All three were central in the writing of the Carbon Neutral case study that appears at the end of each chapter. At the University of New Mexico, James Burbank, Karen Schechner, Shannon McCabe, Andrew Mara, Craig Baehr, and Kristi Stewart helped Richard develop the end-of-chapter questions. Professors Charles Paine and Scott Sanders have offered helpful suggestions to streamline some of the invention techniques described in this book

We also want to thank the many participants in UNM's and Purdue's grant writing and proposal writing workshops over the last 24 years. Their reactions and comments to the ideas shown in this book have helped us sharpen some concepts and strategies. They have allowed us to test these strategies against a wide range of proposals and grants.

Finally, this book is dedicated to Tracey, Emily, Collin, Cathy, Elizabeth, the Thompsons, the Camilo-Apgars, and Jake, who helped us struggle through the rough spots as this book was written and revised for this 3rd edition. In their own ways, they provided the motivation to keep going.

Richard Johnson-Sheehan, Purdue University
Paul Thompson Hunter, Purdue University

WRITING PROPOSALS AND GRANTS

1 AN INTRODUCTION TO PROPOSALS AND GRANTS

Why Write Proposals?

A proposal is a tool for managing change. We write proposals because the world around us is always evolving and changing, creating new problems and new opportunities. We write proposals because, as the old proverb says, "the only constant in life is change." We're always in the process of rebuilding ourselves, our companies, and our organizations. Even the most successful plans, the greatest buildings, and the strongest relationships need to be reimagined, reconsidered, and rebuilt to keep up with a world in flux. Proposals are instruments for managing those changes.

Some people resist change. They worry about losing what they have, so they try to ignore the evolving world around them. Sometimes, they want to hold back change or slow it down. As a writer of proposals and grants, you should view change as your ally, not your enemy. You should recognize that an always-evolving world creates new openings and new opportunities that can be used for starting new projects, growing your business, advancing your nonprofit organization, or doing groundbreaking research. You should look forward to change with optimism. Proposals and grants are tools for steering and shaping your reality, riding the currents of change toward bigger and better things.

This book helps you write effective proposals and grants by using time-tested analysis and persuasion strategies to find new opportunities and solve problems. You will learn how to develop plans for taking action, organize your ideas, improve the clarity of your writing, and persuade your readers to say, "yes."

You don't need to be told how important proposals are to your career or your business. You probably picked up this book because you have an important proposal or grant to write. Perhaps proposals already play a significant role in your work life. Or you are taking a course about writing proposals and grants. In any event, we don't need to convince you that proposals are important tools in the workplace, whether you are a CEO pitching a multimillion-dollar project, a team leader pitching a new product to management, or a researcher seeking funding for new green energy technology.

If you need to write grant proposals, you already know that securing funding is becoming increasingly difficult as grants from government and private funding sources become more competitive. This book will help you take a more business-oriented approach to grant writing, which will strengthen your ability to win funding.

The Power of Persuasion

Have you ever wondered why some people are more persuasive than others? Do they have a knack for changing people's minds or motivating them? Are they more attractive in some way? Do they have a unique way of talking? Are they just lucky?

Persuasive people have usually figured out how to use time-tested rhetorical strategies to influence others. They may have learned them in college or from mentors. Some of them have just noticed that saying and doing things a particular way gets others to say yes. You can learn these rhetorical strategies, too.

This book is based on rhetorical principles for building consensus and persuading others. Rhetoric is the ancient art of persuasive communication, a rich discipline with more than two thousand years of tradition. Rhetoric is often defined as the "art of persuasion," but it is also the shaping of change. Rhetoricians see change as the normal condition of reality, exploring what could be, what might be, or what should be. Rhetoric is a tool for gaining power in an uncertain reality and finding the available means of persuasion within that uncertainty.

Interpreting an Evolving Reality

A big part of persuading others is to accurately *interpret* what is really happening. In this book, you will learn a series of rhetorical strategies for asking the right questions and then imposing cognitive frames on situations that are evolving, uncertain, or chaotic.

Persuasive writers know from experience that they need to first properly interpret what is happening around them by using the information available. Only then will they be able to grasp the big picture by recognizing what information they already have and what information they still need to collect before taking action. Persuasive writers use interpretive strategies to impose order on the current situation. In this book, you will learn these strategies.

Expressing Your Ideas Persuasively

Expression is the performance side of rhetoric. Once you have properly interpreted the current situation, you can develop what proposal writers call a "capture strategy" that will help you to express your ideas persuasively to the readers. You can frame your ideas in powerful and influential ways, using your readers' psychological leanings to weave emotional and authoritative themes into the content and style of your proposal. You will also learn how to use principles of visual design to deliver a professional package that will set your proposal apart from its competitors. Expression is not simply a means for spinning the facts to your advantage. How you say something is what you say.

In this book, you will find this interpretation/expression balance used in each chapter to help you, first, use cognitive frames to impose order on chaotic writing situations and, second, use powerful persuasion strategies to influence your readers. You might even find that the rhetorical strategies you learn in this book are useful far beyond writing proposals and grants. They are tools for analytical thinking, and you can use them to be more persuasive and influential in all aspects of your life.

What Is a Proposal Anyway?

A proposal is a document or presentation that puts a project, product, or service forward for consideration. Typically, a proposal is designed to help the readers solve a problem or take advantage of an opportunity. Maybe they want to add more automation to their manufacturing facility. Maybe they want to figure out how to reduce child poverty. Maybe they see a new way to use artificial intelligence to improve healthcare.

A Proposal Is a Genre

Proposals are a *genre*, which means they tend to follow a common pattern. A genre is a way writers get things done with words and images. For example, let's say you are searching the Internet for reviews of a new movie. As a reader, you already have a good idea about the kinds of information a movie review will have (its content). You also know how that information will be arranged (its organization), what voice or tone the author will probably use (its style), and how the document will look (its design). In other words, both you and the authors know how the movie review genre works, which makes reviews easier to write and read.

The same is true of the proposal as a genre. A proposal typically includes the following major sections (Figure 1.1), not necessarily in this order and not necessarily with these headings:

Introduction—explains the topic, purpose, and main point of the proposal while offering some background information on the topic.

Background, Narrative, or Current Situation—an analysis or description of the problem that the proposal is trying to solve, as well as its causes and effects.

Project Plan or Methods—a step-by-step plan for solving the problem that explains why it would be solved that way and identifies the deliverables.

Qualifications—an overview of the people who would work on the project, the equipment and facilities used, and similar projects that have been successfully completed before.

Conclusion with the Costs and Benefits—a summary of the costs and benefits of the project, a look to the future, and contact information.

Budget and Budget Narrative—A line-by-line breakdown of the project's costs and an explanation of how those costs will allow you to achieve your plan's objectives.

Proposals come in all shapes and sizes, but they usually have these sections, whether they are business-to-business proposals, internal planning proposals, or grant proposals. The proposal genre is designed to answer the following questions:

- What problem or opportunity needs to be solved or pursued, and what's causing it?
- What objectives need to be met to solve the problem or take advantage of the opportunity?
- What is a good plan for solving the problem or capitalizing on the opportunity?
- Why is your company or organization best qualified to do this project?
- How much will the project, product, or service cost?
- What are the tangible benefits of the project, product, or service to the proposal's readers?

These questions might make proposals sound complicated, but if you think about it, you use these questions to make decisions all the time. Aren't these the same questions you would expect a car mechanic to answer before fixing a problem with your car? Wouldn't you expect a surgeon to answer these kinds of questions before she puts you on the operating table? Would you want to do business with a mechanic or surgeon who couldn't answer these questions up front? Of course you wouldn't. Just like you, your readers expect your proposal to anticipate and answer specific questions about the project.

That said, the proposal genre is not a formula into which writers plug content. The pattern shown in Figure 1.1 offers a model to start with, but it's not a template that proposals need to follow. Once you learn how to use the proposal genre, you will find that you can shape this pattern to suit each unique proposal writing situation.

Figure 1.1: The Major Sections of a Typical Proposal

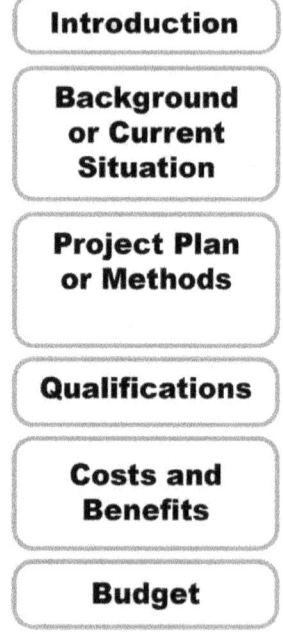

Categories of Proposals

Proposal writers also tend to sort proposals into categories, depending on how the proposal originated and how it will be used:

- **External proposals** are used to conduct business between companies or organizations. These documents tend to be formal because they often serve as interim contracts if accepted by the customer or client. In other words, if an external proposal is accepted, the bidder cannot go back and alter the terms without negotiating those changes.
- **Internal proposals** are used within a company or organization to pitch new ideas, new methods, or new products and services to the company's management. Internal proposals are also used as planning tools within companies by teams of employees.

Proposals are also labeled "solicited" or "unsolicited," depending on who initiated them:

- **Solicited proposals** are requested by customers, clients, or funding sources. Solicited proposals are often initiated through a Request for Proposals (RFP), which might also be called a Call for Bids (CFB), Call for Proposals (CFP), Request for Quotation (RFQ), Request for Information (RFI), or Request for Application (RFA). You will learn more about RFPs in Chapter 2.
- **Unsolicited proposals** tend to be initiated by the person or company who is selling a product, service, or new concept. Companies and consultants will often use unsolicited proposals as sales tools to pitch new products or services to other companies or organizations. Unsolicited proposals are also used internally within companies and organizations to propose changes to policies, methods, and procedures.

These labels mostly help proposal writers keep track of what kind of proposal is being written. An external solicited proposal will be treated with a heightened sense of care (and anxiety) because it was requested by the customer or client, and it will be used outside the writer's company or organization. A bit differently, an internal unsolicited proposal needs to be well written, of course, but the readers are likely more familiar with the problem that the proposal is addressing. So, the proposal might be less formal (and less anxiety provoking).

The Proposal-Writing Process

In this book, you will learn a process for writing proposals that includes five stages:

> **Stage 1: Planning and Research**—developing your ideas, collecting information, and identifying persuasive strategies that will help you convince the readers to say yes to your project

> **Stage 2: Organizing and Drafting**—arranging your ideas and writing a rough draft of the proposal

Stage 3: Improving the Style—putting your ideas in plain language and strengthening your persuasiveness by appealing to your readers' motives, values, attitudes, and emotions

Stage 4: Designing the Document—using page design and graphics to reinforce your argument and to make it more attractive, accessible, and usable for the readers

Stage 5: Revising and Editing—refining and strengthening the organization, style, and design of the draft.

Proposal writers and grant writers rarely follow this process in lockstep. Instead, they move back and forth among the stages as they draft each part of the proposal separately.

Where Does Artificial Intelligence Fit In?

Artificial Intelligence (AI) applications are valuable tools that can help you write proposals and grants. AI applications are especially helpful when doing research, outlining parts of the document, designing how the document will look, and editing to make it easier to read and grammatically correct.

With these capabilities alone, AI has already accelerated the proposal writing process and helped writers produce easier-to-read proposals that have few, if any, mechanical and grammatical errors. But you won't be able to use AI applications to completely write proposals and grants for a couple of important reasons. First, proposals are written about the future, which can be difficult for AI applications to envision. Writing about the future requires creative thinking and innovation. AI applications are better at collecting and aggregating existing information about the past or present.

Second, AI applications can struggle to predict how living, breathing human beings will interpret a real situation and what they will say yes to. This limitation of AI is important because humans are your customers, clients, and funders who will need to be motivated to accept the proposal or grant. As we will show you in this book, readers tend to frame issues through their own values and emotions, and they rarely rely exclusively on reasoning.

Believe it or not, a completely reasonable or logical proposal may not win readers over, but a proposal that connects with your readers' values and feelings can be highly persuasive. AI applications may be able to figure out the general sentiment of a text, but they cannot meaningfully empathize with your proposal's readers or account for their lived experiences.

Throughout this book, we will highlight places where AI applications can be helpful for doing research, outlining, designing, editing, and proofreading. In places where AI applications are not helpful, we will show you how to use your human creativity to generate new ideas and winning strategies that an AI application may miss. AI is certainly a major technological development, but writing proposals and grants is fundamentally a matter of human beings persuading other human beings to say yes.

Grant Proposals

If you need to write grants, you're probably wondering how writing a grant proposal compares to writing a business proposal. Don't worry, grants and business proposals are actually quite similar.

Keep in mind that grant proposals are essentially business proposals. When you submit a grant proposal, you are promising to do a project that the funding source wants completed. Funding sources, even the U.S. Federal Government, are not giving away money because they are generous benefactors—well, some are. But most are not. Funders have their own goals and motives, and they only fund people and organizations who will help them reach their objectives.

Let's imagine you have $100 million to start your own foundation. Imagine you can give this money to any worthy cause you want—other than yourself, of course! How would you decide which causes to give your money to? More than likely, you would begin by identifying a few types of social problems that you want your foundation's funding to solve (e.g., poverty in rural areas, declining support for the arts, serious transmittable diseases). Your foundation will offer funding to organizations that could do that kind of work on your behalf.

But you're not going to give your money to just anyone. Your foundation will only give its money to people who can help you reach your objectives. In other words, those organizations will be providing a service for your foundation's money, much as any business provides a service for another business.

What about writing grants to the federal government, such as proposals to the National Science Foundation (NSF), the National Institutes of Health (NIH), the National Endowment for the Humanities (NEH), or the Department of Defense (DoD)? Keep in mind that these funding agencies are not simply looking to give away taxpayers' money. Quite the opposite is true.

These foundations only fund projects that help the federal government achieve its goals. To illustrate, let's say you are a medical or healthcare researcher writing a grant proposal to the NIH. Your application would be much more attractive to the NIH Board of Directors if it addresses a recognized problem in our society (e.g., cancer, deadly viruses, bioterrorism) or your community (e.g., addiction, low birth weight, farm safety). Your proposal will also be more desirable if it shows how your project moves our society one step further toward solving a problem that is important to the federal government.

Foundations, governments, and corporations offer grant funding because they have identified important problems that they would like their money to solve. Even the most altruistic foundations have motives for offering their money, and they will want you to show that your project will meet their goals, that you and your team can complete the project, and that the project will be on time and under budget

In the end, grant proposals are a special kind of business proposal. As a grant writer, you should keep in mind that you always need to consider the business side of research when pitching a project. The sooner you start thinking about grant proposals as business proposals, the sooner you will be successful in securing funding.

Chapter Summary and Looking Ahead

Proposals and grants are written to manage change. To accomplish your goals as a proposal writer or grant writer, you can use rhetorical strategies to impose intellectual order on chaotic situations. You can then invent the content of your proposal, organize your argument, use style to enhance your voice and tone, and deliver a professional document. This book will show you how to use rhetorical strategies to write proposals that succeed. Rhetorical strategies have been used for millennia for one simple reason: they work!

Try This Out!

1. What are some of the problems or opportunities on your campus, in your workplace, or in your community that might be addressed with proposals or grants? List out a few of these problems/opportunities. Under each of these topics, offer some reasons why these problems/opportunities might exist. Then, write down some ways in which you might use proposals to address them.

2. Find a proposal or grant on the Internet, in your local library, or at your workplace. Write an email to your instructor in which you study the strengths and weaknesses of the proposal's content, organization, style, and design. Consider the following questions:

    a. Does the proposal include the major sections of the proposal genre described in this chapter?
    b. Does the proposal include any sections that go beyond the proposal genre discussed in this chapter?
    c. Are there any sections missing that you would expect? If so, explain why you think the proposal writer chose to leave out specific sections.
    d. Do you think the writer used AI to produce the proposal? Why or why not?
    e. Overall, do you think the proposal is effective or ineffective? Why?

3. Using social media, find a person who writes proposals or grants regularly. Reach out to this person and ask what kinds of proposals they write. Ask what kinds of situations require a proposal to be written. Present your findings to your class, highlighting any advice that the proposal writer or grant writer offered to you about writing effective proposals.

4. Name some situations on your campus, in your workplace, or in a society where rhetoric is used to manage change. How is persuasion used to first "identify the available means of persuasion" and then alter that situation to the advantage of the speaker or writer? Who tends to use persuasion in our society? How do they use it?

5. Imagine you have just been assigned to an advisory committee that has been asked to develop ways to reduce the number of cars being driven to your campus or workplace. What information do you already have on this topic? What information would you still need to

solve the problem? What are some questions you would need to be answered before you and your team begin writing a proposal to solve this problem?

Case Study: The Carbon Neutral Campus Project—Introductions

To illustrate the proposal writing process, each chapter in this book will include a scene from an ongoing case study. This case study will show how the rhetorical strategies discussed in each chapter of *Writing Proposals and Grants* can be used to develop a proposal.

This case study is set on the campus of the fictional Durango University, a mid-sized private university with 12,000 students. The university was founded in 1912 on an eighty-six-acre campus in southwestern Colorado. Due to environmental concerns and soaring energy costs, the university's president, James Wilson, recently announced that one of his strategic goals is to convert the century-old campus into a "green campus" that would eliminate or offset its emissions of greenhouse gases like carbon dioxide, methane, and nitrous oxide.

President Wilson explained that energy costs were severely harming the university's budget, and these costs would likely go much higher as the United States confronted issues involved in the climate crisis. So, he wanted to begin converting Durango University to a "net-zero carbon" campus. He called this initiative the "Carbon Neutral Campus Project."

Like most campuses built at the turn of the 20th century, Durango University's campus has several aging buildings and an incredibly inefficient central heating and cooling system. Converting the campus to renewable forms of energy would not be easy.

Nevertheless, the conversion will need to happen at some point. Plus, as President Wilson pointed out, becoming a green campus will be an attraction to students and potential faculty around the world. Durango University is already an attractive place to study and teach because of its location in the mountains. Transforming the university into a green campus would make it an even more attractive place.

President Wilson delegated the Carbon Neutral Campus Project to Anne Hinton, the Vice President for Physical Facilities. She began assembling a team of people who could help her write a grant proposal to fund the project. She chose the following people to participate:

- George Tillman, a professor of environmental engineering who researches renewable energy systems, especially geothermal power plants.
- Calvin Jackson, a local contractor who specializes in renovating buildings to make them energy efficient.
- Karen Briggs, a professor of social science who specializes in human resource issues.
- Tim Boyle, a student chair of the Student Environmental Council, a campus organization that advocates for environmental issues.

The five members of the task force will need to bring their strengths, concerns, and opinions to the project. All five of them recognize that this project is complex and potentially controversial. Where should they start? How should they proceed with writing the proposal?

In each chapter of this book, you will see how Anne, George, Calvin, Karen, and Tim use the proposal-writing process described in this book to help them write a grant that will capture funding for the Carbon Neutral Campus Project.

2 REQUESTS FOR PROPOSALS (RFPs)

Two Basic Reasons for Writing Proposals

A request for proposals (RFP) is a public announcement that a company or government agency is putting a project or opportunity out for bid. Typically, an RFP will describe the project or opportunity in detail while identifying its objectives, specifications, and deliverables. In some cases, RFPs will identify what kinds of companies or organizations are eligible to bid for the project.

Similarly, grant opportunities usually begin with an RFP, sometimes called a Request for Application (RFA) or a Call for Proposals (CFP). Grant RFPs are similar to corporate and government RFPs but tend to be more open-ended. A grant funding source, whether a government agency or a private foundation, will use the RFP to identify the objectives that any fundable project would need to meet and describe who is eligible to apply for the funding. Grant RFPs tend to be more open-ended than business RPFs because funding sources want to consider a range of potential projects that align with the objectives identified in the RFP.

RFPs can be found in various places:

- Corporations usually send their RFPs directly to past bidders, or they place RFPs on their websites and social media sites where search engines can find them. RFP databases like FindRFP.com and the RFPDatabase (rfpdb.com) can be used to search for opportunities.
- The U.S. government publishes its RFPs on the SAM.gov website. The SAM.gov search engine will allow you to use keywords to find grant opportunities.
- State and local governments often publish RFPs on their websites and occasionally in the classified ads sections of major local newspapers.
- For grant funding opportunities, most government agencies and private foundations list their RFPs on Grants.gov or the Pivot-RP database (pivot.proquest.com). These databases are keyword searchable.

RFP databases tend to offer a synopsis or abstract of the full-text version of the RFP. If the synopsis interests you, you can download the full version and sometimes an application packet.

When reading through RFPs, you should keep in mind that proposals and grants are usually "requested" for two basic reasons: 1) a problem has emerged that needs to be solved or 2) an opportunity has appeared that calls for some kind of action. So, as you read each RFP, keep asking yourself, "What has changed recently that created this problem or opened up a new opportunity?" For example, let's say a company or federal agency sends out an RFP that's requesting bidders who can upgrade the electrical wiring in an office building. Bidders should ask themselves, "What changed?" Is the current wiring failing, becoming obsolete, or falling out of code? If so, that's a problem. Or does the company or federal agency see a potential future use for the building that they want to take advantage of? That's an opportunity.

Grant writers ask similar questions about change. If a state agency is offering grants for harm-reduction programs for teens who use illicit drugs in rural areas, grant writers will ask, "What's changed recently that makes this problem especially important right now?" Has there been a sharp increase in drug use in rural areas? Have economic conditions in rural parts of the state worsened in some way? Are teens self-medicating with illicit drugs because of changes in healthcare policies that restrict their access to therapy or prescribed drugs? After all, illicit drug use among teens is not a new problem. So, what changed that spurred a state agency to send out this RFP now?

In this chapter, you will learn how to read RFPs carefully and rhetorically. You will learn how to analyze the problem your customer, client, or funding source is trying to solve or the opportunity they want to take advantage of.

Reading RFPs

RFPs are called by a variety of names, depending on the kind of work or information the customer or client is looking for. An RFP might be referred to as a Call for Proposals (CFP), a Request for Application (RFA), an Information for Bid (IFB), a Request for Grant Proposals (RGP), a Call for Quotes (CFQ), or an Advertisement for Bids (AFB), among other names. Whatever it is called, an RFP is essentially an announcement that a customer, client, or funding source is seeking proposals for a specific project or opportunity.

Some RFPs are brief, using only a few hundred words to describe the project and state the deadlines. Other RFPs, especially those from government agencies, can run dozens of pages.

Two Example RFPs

Figure 2.1 shows an RFP synopsis that was sent out by the City of Fitchburg in Wisconsin. The full RFP would be too long to print here, but the synopsis offers a good overview of the project that is being put out for bid. In this RFP, the City of Fitchburg is looking for a planning or design firm that can offer the following services:

> The purpose of this plan is to create a neighborhood plan and development strategy to meet the City of Fitchburg's land use needs and position the city to submit applications for USA amendments to CARPC and the Wisconsin D.R. to accommodate growth.

The rest of this RFP synopsis offers background information on the project and describes the city's goals and interests.

Figure 2.1: Request for Proposals for Neighborhood Planning

RFP - Greenfield Neighborhood Plan

Request Category Code: 906-57 Land Devel/Planning Arch, Planning/Land Use, 925-61 Land Devel/Planning/Engineer Description: This request for proposals (RFP) is issued by the City of Fitchburg for the purpose of securing a qualified planning or design firm to develop a neighborhood plan for the approximately 975-acre Greenfield Neighborhood. The plan development will include research and site analysis related to economic development positioning, residential capabilities, and market considerations. The plan development process will include landowner and stakeholder interviews, intergovernmental coordination, and public involvement. The plan will be considered for adoption as part of Appendix A of the City's Comprehensive Plan. Located in the east-central portion of the City of Fitchburg, this neighborhood is located south of the existing urban service area.

The purpose of this plan is to create a neighborhood plan and development strategy to meet City of Fitchburg land use needs and position the City to submit applications for USA amendments to CARPC and the Wisconsin DNR to accommodate growth. The goals of the neighborhood plans are to develop balanced neighborhoods that will integrate compatible land uses, provide economic opportunities, and increase the municipal tax base, while balancing transportation and service needs, minimizing environmental impacts, and recognizing the variety of concerns and issues that may arise from the planning process. This plan is necessary to evaluate the expansion of the central urban service area (CUSA). The Plan is to be used to determine which areas of the Greenfield Future Urban Development Area (FUDA) are suitable for urban expansion. The City of Fitchburg is working closely with the Capital Area Regional Planning Commission (CARPC) on a pilot process to better integrate neighborhood planning and the Urban Service Area amendments. As such, this plan is expected to also be used as the application(s) for urban service area amendments to CARPC. The goal of this project is to create a neighborhood plan that will guide future development. As such the plan should be thorough enough in the background data, evaluation, analysis, and recommendations that if a developer proposes a project that reflects the plan, it should be able to be approved and built with minimal changes.

The RFP shown in Figure 2.2 is quite different. This RFP from the National Science Foundation (NSF) is soliciting grant proposals for research on environmental sustainability. The actual RFP (P.D. 23-7643) is the length of a small book. In this excerpt, you will notice that the request is more open-ended than the one in Figure 2.1. The NSF is looking to fund projects that "promote sustainable engineered systems that support human well-being and that are also compatible with sustaining natural (environmental) systems." In other words, a broad range of possible projects could fit this grant funding opportunity.

In the RFP, the NSF offers some guidance about how they think the objective can be achieved, but they also leave a great amount of flexibility about the kinds of research projects that would be appropriate.

Figure 2.2: An RFP for a Federal Grant

Environmental Sustainability

 National Science Foundation

The **Environmental Sustainability** program is part of the **Environmental Engineering and Sustainability** cluster together with 1) the **Environmental Engineering** program and 2) the **Nanoscale Interactions** program.

The goal of the **Environmental Sustainability** program is to promote sustainable engineered systems that support human well-being and that are also compatible with sustaining natural (environmental) systems. These systems provide ecological services vital for human survival. Research efforts supported by the program typically consider long time horizons and may incorporate contributions from the social sciences and ethics. The program supports engineering research that seeks to balance society's need to provide ecological protection and maintain stable economic conditions.

There are five principal general research areas that are supported.

- **Circular Bioeconomy Engineering:** This area includes research that enables sustainable societal use of food, energy, water, nitrogen, phosphorus, and materials, with the reduction and eventual elimination of fossil fuel combustion that lacks carbon capture. The program encourages research that helps build the raw material basis for the functioning of society principally on biomass, drawing heavily on sustainable agriculture and forestry. Additionally, material flows must reduce or preferably eliminate waste, with an emphasis on closed-loop or "circular" processing.
- **Industrial ecology:** Topics of interest include advancements in modeling such as life cycle assessment, materials flow analysis, net energy analysis, input/output economic models, and novel metrics for measuring sustainable systems. Innovations in industrial ecology are encouraged.
- **Green engineering:** Research is encouraged to advance the sustainability of manufacturing processes, green buildings, and infrastructure. Many programs in the Engineering Directorate support research in environmentally benign manufacturing or chemical processes. The Environmental Sustainability program supports research that would affect more than one chemical or manufacturing process or that takes a systems or holistic approach to green engineering for infrastructure or green buildings. Improvements in distribution and collection systems that will advance smart growth strategies and ameliorate effects of growth are research areas that are supported by Environmental Sustainability. Innovations in management of storm water, recycling and reuse of drinking water, and other green engineering techniques to support sustainability may also be fruitful areas for research.
- **Ecological engineering:** Proposals should focus on the engineering aspects of restoring ecological function to natural systems. Engineering research in the enhancement of natural capital to foster sustainable development is encouraged.
- **Earth systems engineering:** Earth systems engineering considers aspects of large-scale engineering research that involve mitigation of greenhouse gas emissions, adaptation to climate change, and other global concerns.

All proposed research should be driven by engineering principles and be presented explicitly in an environmental sustainability context. Proposals should include involvement in engineering research of at least one graduate student, as well as undergraduates. Incorporation of aspects of social, behavioral, and economic sciences is welcomed.

Interpreting an RFP

The ability to properly interpret RFPs is a valuable skill, whether you are writing a business proposal or a grant proposal. Larger companies will often hire an RFP manager who coordinates the routing of RFPs within the organization. Similarly, nonprofit organizations hire development officers who search for grant opportunities and stay in close contact with government and private foundations. These professionals look for proposal opportunities by monitoring SAM.gov, Grants.gov, trade periodicals, RFP databases, and incoming emails. Then, the RFP manager or development officer coordinates between potential funding sources and the teams who are writing the proposals or grants. This ensures that the RFP guidelines are followed properly and that any amendments to the RFP reach the right people.

In some special cases, RFP managers and development officers may have the chance to provide feedback to customers, clients, or funding sources on drafts of RFPs. In these situations, a draft of an RFP would be released to solicit feedback from companies or organizations that are likely to pursue the opportunity. By providing comments on the draft, an RFP manager or development officer can help refine the advertisement, often to the benefit of their company or organization.

One thing to keep in mind, however, is that sometimes the best proposal is no proposal at all. If your company or organization cannot meet the RFP's requirements or if the funding source is requesting something different from what you offer, you can save yourself time and effort by looking for an RFP that fits what your company or organization does. Unless your company or organization can do almost exactly what the funding source is asking for, it's usually a good idea to spend your time and efforts on a different proposal or grant. But if the RFP seems in line with what you or your organization can provide, you should begin figuring out the status of the opportunity.

Determining the Stasis of an Opportunity

Whether you are responding to an RFP or writing an unsolicited proposal, determining the status of the opportunity is an important first step. In rhetoric, we call this the *stasis*, where both sides agree to a discussion. In Greek, *stasis* refers to a state of balance or stability. Before negotiation can take place, the two sides need to agree on what is being negotiated. By figuring out the stasis of an opportunity (i.e., what you are negotiating about), you are establishing a baseline understanding of the problem/opportunity with the customer, client, or funding source.

Using the Five-W and How Questions

The first step in figuring out the stasis of a problem or opportunity is to analyze the elements of the rhetorical situation. Most people are familiar with the who, what, where, when, why, and how method that journalists use to develop a news story. Proposal writers can use this journalistic method to sort out the elements needed to understand the current situation:

- *Who* exactly are the readers, and who else might be involved?
- *What* do the readers need?
- *Where* is the work site? Where do we need to submit the proposal?
- *When* are the deadlines for the proposal, and when does the project need to be completed?
- *Why* is the customer, client, or funding source looking for someone to do this project?
- *How* should the project be completed?

Even simple answers to these questions will provide a basic understanding of the status of the problem/opportunity for which the proposal is being written.

You will notice that the why and how questions are rarely answered directly in an RFP. For example, in the RFP in Figure 2.1, the City of Fitchburg never really tells us why it wants a plan to redevelop the Greenfield neighborhood in their town. The RFP only gives us a few ideas about what they have in mind. As proposal writers, we can speculate about answers to the why and how questions: Did Fitchburg recently annex this area? Has this neighborhood experienced rising crime rates? Are they trying to build economic opportunities? Was a bill passed in the Wisconsin State Legislature to make funds available for urban planning and development?

Of course, knowing *why* the RFP was written and *how* the customer or client wants the project completed will improve a proposal's chances of winning this contract. And yet, the RFP from Fitchburg only gives us hints about why they are asking for bidders to do this project and how they want it to be completed. Later in this chapter, we will discuss how to handle these kinds of why and how questions.

Using Proposal Opportunity Worksheets

When interpreting an RFP, you can start out by identifying answers to the who, what, where, and when questions that can be found in the advertisement. Then, you can make some guesses that tentatively answer the why and how questions. When you are finished answering the Five-W and How questions, you will have made a good start toward understanding what the customer, client, or funding source is asking for.

Proposal writers and RFP managers will often use a worksheet to help them sort out the elements of a proposal-writing situation. A Five-W and How Worksheet, like the one shown in Figure 2.3, can help you break down the RFP into five key questions. As you read through an RFP, you can identify places where information related to the who, what, where, when, why, and how are mentioned. Then, make some guesses about how you might answer the why question.

Many proposal writers and RFP managers also like to fill out an RFP Analysis Worksheet like the one shown in Figure 2.4. This kind of worksheet allows them to break down the RFP into categories that highlight the key parts of the proposal or grant opportunity.

RFP managers often forward these kinds of RFP Analysis Worksheets to teams in the company or organization to alert them to business or funding opportunities. This kind of worksheet summarizes the Five-W and How answers in an easily digestible format, so any interested teams can decide quickly whether an opportunity is something they might want to pursue.

Figure 2.3: Filling Out the Five-W and How Worksheet

Five-W and How Worksheet

Who?
What?
Where?
When?
How?
Why?

Defining the Problem or Opportunity

Consultants will often say, "There are no problems, just opportunities." And, in the optimistic world of consultant-speak, that's true—a problem is just an opportunity to make things better.

In proposal and grant writing, the word *problem* is typically used in a forward-thinking way. Proposals are written to solve problems, so figuring out exactly what the problem is tends to be a common first step. The word problem lends a sense of urgency and importance to a project. So, "problem-solving" is usually viewed as a positive, action-oriented way to look at the proposal-writing process.

A successful proposal begins with a clear understanding of what is causing the problem. So, you should first use the clues offered by the customer, client, or funding source, and your own research to start figuring out *why* the problem exists. To help you answer this why question, you can use a discovery tool from rhetoric that's called *stasis questions*. Usually, stasis is determined by answering four questions:

1. Is there a problem?

2. What exactly is the problem?

3. How serious is the problem?

4. What kind of proposal would solve the problem?

As soon as you have an answer to each of these questions, you will have a clearer notion of how to start the proposal-writing process.

Step 1: Is There a Problem?

This first question is about whether a problem exists. This might seem a bit odd. After all, why would your customer, client, or your company's executives solicit proposals for a problem that doesn't even exist?

Sometimes, people worry that they have a problem when they sense that change is happening. Maybe they are seeing a slight dip in business, so they get a little nervous and begin asking, "What's wrong? What should we do about this?" Maybe nothing is wrong.

If your customers, clients, executives, or you aren't even sure if a problem exists, your best move might be to first recommend a research study that determines whether a change or a dip in sales is just a natural fluctuation in the market.

Figure 2.4: RFP Analysis Worksheet

RFP Analysis Worksheet

Project Title: Solicitation Number: Date Advertised or Received:	
Client: Point of Contact: Deadline for Proposal Submission: Address for Proposal Submission:	
Summary of Proposal Opportunity	
Comments and Recommendations	
Accept or Reject	

Reviewer: Phone Number: e-mail address:	Reviewer Initials	Date Reviewed

But wait. Business is business, and the customer is always right. Right? Another old saying among consultants is, "You can only sell an empty box once." In other words, you should write only proposals that solve real problems. Two bad things can happen when you write a proposal to solve a problem that doesn't exist: a) while reading your proposal, the readers see that they don't really have a problem, so you just wasted your time doing all that work, or b) they award you the project but later realize you have sold them something that they didn't need.

You would be better off passing on the opportunity and not harming your relationship with your customer, client, or managers. Sometimes, the best proposal is the proposal that was never written.

Step 2: What Exactly Is the Problem?

All right, you decide that a problem exists. Now, you need to define it. Of course, your first step is to ask your customers or clients to help you figure out the exact problem they are trying to solve. A good technique is to ask them the Five-W and How questions:

- *What* is happening?
- *Who* has already been involved with the problem?
- *Where* is this problem occurring?
- *When* does it happen?
- *Why* do you believe this problem is happening?
- *How* do you think this problem can be solved?

After they answer your Five-W and How questions, follow up by asking them, "OK, what is new or has changed recently that may have caused this problem?" This is a key question. You see, problems almost always emerge because something has changed. This means the problem your customers, clients, or managers have identified is the symptom of something that changed recently.

For example, let's say a school district in an affluent community is having trouble attracting high-quality teachers. The school board wants you to come up with a marketing plan that entices strong teachers to apply for jobs in the district. As you do research on the problem, though, you soon discover that the best teachers left the district because they could not afford to live in the community. When you interview former teachers, they tell you they wanted to stay, but housing is expensive in the area, and the commute from other areas is time-consuming and dangerous. You quickly see that the root problem might not be attracting good teachers: it is retaining good teachers. Of course, your marketing proposal would include a plan to attract strong teachers, but you could also enhance your proposal by addressing the root problem of affordable housing for teachers.

Ask yourself: In an affluent community that is having trouble attracting and retaining high-quality teachers, what has changed in this community that created this problem? Have housing costs increased suddenly? Have property taxes gone up? Are the more experienced teachers retiring because of the commute? By paying attention to these kinds of underlying changes, you can usually find the specific problem that needs to be solved.

The visible problem might be merely a symptom of a deeper change that's happening below the surface. If you can identify what's changing and then properly define the problem, you will have an edge over your competitors.

Step 3: How Serious Is the Problem?

This third stasis question helps you understand the seriousness of the problem. This will help you prioritize what needs to be done first, second, third, and so on. Specifically, you need to figure out the most pressing causes and effects of the problem. What aspects of the problem can wait while higher-priority issues are handled? By deciding what is most serious or pressing, you can prioritize the steps in your plan to focus on the most urgent issues.

A good rule of thumb when determining the seriousness of a problem is to "put first things first." In other words, the most pressing cause of the problem is likely to be the one that needs to be solved first. Once that part of the problem is solved, you can then work on the next most serious cause of the problem. Putting first things first will help you innovate better and build a better plan for solving the overall problem.

Step 4: What Kind of Proposal Is Needed to Solve the Problem?

This final question will help you figure out what kind of proposal fits the project. Proposals tend to fall into the four categories shown in Figure 2.5.

The type of proposal you need to write depends on two things: a) the problem or opportunity you are trying to address and b) the deliverables you are expected to provide your customers or clients with when the project is completed. Deliverables are the tangible objects or services that result from the project, such as the things (e.g., products, reports, plans, data sets) that will be delivered to the readers as the project moves forward and when it is completed.

Let's look more closely at the four types of proposals and their deliverables.

Research Proposals

Research proposals describe methods for gaining insight into a particular problem or opportunity. Scientists use research proposals to request funding or approval to conduct an empirical study or develop a prototype. An electrical engineer might write a research proposal to figure out why an undersea robot shuts down when it reaches five hundred meters below the surface. Meanwhile, a biologist might write a grant proposal to the National Science Foundation (NSF) to study the yearly migration of a sandhill crane population.

A research proposal usually describes a study that will generate data or observations and fill in a knowledge gap. The deliverable for a research proposal is typically a report or article that explains the results of the research. In some cases, a prototype might also be a deliverable.

Research proposals are also written for clients who need to better understand a situation, problem, or opportunity. For instance, a client is struggling with quality control issues in their manufacturing plant, and it's causing them to lose customers. A research proposal would propose a study to figure out what is causing these quality control issues. This type of research proposal would describe the methodology used to gather information (e.g., interviews, surveys, observations, and testing).

The "deliverable" for this kind of proposal would likely be a final report in which the findings are presented and explained. In some situations, research proposals offer recommendations for taking action or making changes.

Planning Proposals

Planning proposals offer step-by-step strategies for making improvements to existing products or services or making significant changes. A planning proposal might be used to describe new manufacturing processes, suggest changes to current business practices, or describe a marketing campaign that can increase sales. In many cases, the planning proposal itself is the deliverable. Consultants will often write planning proposals to describe how a project should be completed. The customer or client would then use the planning proposal to make changes themselves or to hire someone else (perhaps the company that wrote the planning proposal) to develop an implementation plan and do the work.

Implementation Proposals

Implementation proposals are written when the readers already know what they want and already have a plan for doing it. They are seeking someone who can implement the plan. For instance, construction contractors often write implementation proposals that show how they would turn an architect's or designer's drawings (the project plan) into an actual building or a product.

In an implementation proposal, customers and clients are looking for specific timelines, the names of the personnel who will be involved, a list of materials, and an itemization of costs. The deliverables for an implementation proposal are typically the promised final products or services, as well as a "completion report" that documents the implementation process and describes any deviations from the original plan.

Estimate Proposals (Sales Proposals)

Estimate proposals, often referred to as "sales proposals," tend to offer a product or service for a specific cost. In these cases, the customers or clients know what product or service they need. They simply want bidders to tell them how much that service or product will cost and when it can be delivered. Estimate proposals tend to be used to bid on standard services, like customer service, legal representation, janitorial services, maintenance work, or clerical services.

Of course, these four types of proposals overlap, and in many cases two different kinds of proposals might be merged. For instance, a planning proposal might include both a research phase (to identify the causes of the problem) and a planning phase (to describe a solution to that problem). Likewise, an implementation proposal might offer a project plan and provide a general description of how the plan can be implemented. Then, if the customers or clients approve the project, a more detailed implementation proposal would expand that project plan by including details about how exactly the project would be completed.

Figure 2.5: Four Types of Proposals

Type of Proposal	Problem	Purpose	Deliverables
Research Proposal	Needs insight or empirically produced facts	Proposes a research project; often requests funding	Report or publication that describes and analyzes results of the study; might offer recommendations
Planning Proposal	Needs a plan that outlines a general strategy	Proposes to develop a strategic plan for addressing a problem/opportunity	Plan that describes a general strategy for solving the problem or taking advantage of the opportunity
Implementation Proposal	Needs to implement a strategic plan	Offers a detailed plan for implementing a project	Completion of the project and a completion report that demonstrates and measures the results of the project
Estimate Proposal (Sales)	Needs to provide a cost for a product or service	Provides a cost estimate for a product or service	A product or service

Talking to the Point of Contact

Once you have worked through the Five-W and How questions and answered the four stasis questions, you are ready to talk to the Point of Contact (POC) listed on the RFP. Email is usually the best way to make contact, and you can request a meeting via phone call or video conference. Of course, there is no universal script that you can use to talk with a POC. Figure 2.6 shows how you can phrase statements and questions to receive valuable feedback from a POC.

Before meeting with the POC, make sure you have determined the status of the problem or opportunity so you can ask specific and targeted questions. Most POCs will default to "yes" and "no" answers if you don't ask good questions. So, you want to be prepared with informed, specific, and open-ended statements and questions that invite them to elaborate on what was provided in the RFP. If they see you have done your research, POCs will tend to be more helpful, often giving you added insight into the problem the company, client, or funding source is trying to solve.

Even if you do ask open-ended questions, you will often find that POCs are not as forthcoming with additional details as you would like. Many POCs are careful to avoid appearing to tip the scale toward one bidder or applicant—though that does happen. That said, POCs do want to receive the best proposals possible, so they may be willing to tell you what kinds of projects may be "most competitive" for the opportunity. They may even be able to tell you whether your proposal describes the kind of approach they are looking for. So, to get better and more useful answers from a POC, you should ask specific questions based on the research you have already done about their project, their company, and their needs.

Figure 2.6: Talking to the Point of Contact

Comment or Question	Intent of Comment or Question
"Here is our understanding of your current situation. What else should we know?"	Allows you to confirm your answers to the who, what, where, and when questions. Here is also your chance to clarify any uncertainties about the project's details.
"What created the need for this RFP?"	You are trying to find out two things with this question. First, what *changed* at the client's company or organization to create this proposal opportunity? Second, *why* are they asking for proposals?
"Here is our understanding of why you are looking for someone to do this work. What else might you add?"	Allows you to test out your best answer to the why question. Talk about how changes in the current situation brought about this opportunity for the customer or client. Also, in most cases, phrase your answer to the why question as an "opportunity," not a problem.
"Besides the deliverables mentioned in the RFP, what other kinds of deliverables do you expect when the project is complete?"	This question has two purposes. First, the POC's answer may give you an idea about what the customers or clients expect bidders to propose. Second, the answer should tell you what kinds of documentation the readers expect when the work is finished.

Comment or Question	Intent of Comment or Question
"What other sources of information might we access to help us write a proposal that fits your needs?"	In some cases, the POC will have more information available if you ask. Also, published reports or websites may be available that refer to their current situation.

POCs for federal and state government grants will often be more helpful depending on what agency is offering the grant opportunity. As soon as you know what kind of project you are going to propose, a POC, usually called a "program manager," may be able to coach you on ways you can make the project more attractive to the reviewers. However, you must enter the conversation with a solid sense of your project and what it will accomplish. That way, the POC can tell you what has worked in the past and perhaps what would be attractive right now.

Chapter Summary and Looking Ahead

Ultimately, you and your readers need to agree about the nature of the problem/opportunity and the kinds of projects that would solve the problem. You need to agree about the who, what, where, and when. And you need to do research to figure out the why and the how. In other words, you need to agree that there is a problem, what the problem is, and what kind of proposal is needed to solve that problem.

Only then will you be able to write an effective proposal. In this chapter, you learned how to interpret RFPs and two important steps for figuring out the stasis of the proposal writing opportunity (i.e., the Five-W and How questions and the four stasis questions). In the next chapter, we will build on this understanding of stasis by defining the rhetorical situation in which you are submitting the proposal.

Try This Out!

1. Find an RFP in SAM.gov, Grants.gov, a grant RFP database, or the newspaper classifieds. Write a memo to your instructor in which you use the Five-W and How questions to summarize the RFP. Then, discuss why you think the customer or client is looking for someone to do the work described in the RFP. Specifically, discuss what might have "changed" to create this proposal opportunity. And finally, discuss some projects that might be suitable for this RFP.

2. Analyze an RFP in your field or area of interest by using the RFP Analysis Worksheet shown in Figure 2.4 of this chapter. According to the worksheet, what kinds of information would you still need to gather if you were to respond to this RFP? What questions

would you need to ask the Point of Contact to clarify what kind of work is needed and why it is needed?

3. Research a problem or opportunity on your campus, in your workplace, or in your community, such as parking, health care, or safety. What has changed recently and brought about these problems/ opportunities? What are the underlying problems that created these problems? Write a memorandum to your instructor in which you discuss how change has created these problems/ opportunities. In your memo, make some guesses about the reasons why these problems/ opportunities have not been addressed yet.

4. Apply the four stasis theory questions to a problem on your campus, in your workplace, or in your community. Is there really a problem that can be addressed? What exactly is the problem? How serious is the problem? What kind of proposal (research, planning, implementation, or estimate) would be needed to address this problem?

5. Call a Point of Contact listed on an RFP. Politely tell the POC that you are learning how to write proposals and grants. Then, ask permission if you can interview the POC, inquiring about what kinds of questions might be appropriate for a proposal writer to ask. Ask what answers the POC can give to proposal writers who call. Report your findings to your class.

6. With a team, analyze the following RFP using the Five-W and How questions and the four stasis questions.

> RFP: Campus Safety Assessment SOL 45-9326. DUE 10/10/24. POC James Sanchez, Assoc. V.P./Student Affairs (318) 555-0103. Bentworth University, a research university serving more than fifteen thousand on-campus and commuter students and located near downtown Bentworth is soliciting proposals for an assessment of safety on campus. The objective of such an assessment would be to determine the causes of a recent increase in reported crime on campus. We are especially interested in addressing forms of crime like assault, graffiti, and theft. Depending on the report's outcome, special priority will be given to consultants who can also help develop a plan to reduce crime on campus. Also, special consideration will be given to proposals that offer non-intrusive methods for collecting data and information. Interested parties should submit a five-to-seven-page pre-proposal that offers a general sense of how they would go about assessing crime on campus. The Student Affairs office will select five parties to submit full proposals. At that point, more information will be offered. The due date for pre-proposals is October 10, 2024. Full proposals will be due on December 1, 2024.

Then, answer these questions: What might be the problem underlying the current situation? What might have changed to create this opportunity? How serious is the problem, and what are its most urgent aspects that need to be dealt with first? What kind of proposal is the client looking for the bidders to write? What information do you still need to write the proposal? What are some questions you would need to ask the Point of Contact?

Case Study: The Carbon Neutral Campus Project—What Is the Problem?

At their first meeting in the Durango University Student Union, Anne Hinton, George Tillman, Calvin Jackson, Karen Briggs, and Tim Boyle began laying the groundwork for the Carbon Neutral Campus Project. They were all excited about the project and were looking forward to working with each other.

After introducing themselves, they spent some time discussing their backgrounds and their expectations for the project. They looked over a Request for Proposals (RFP) from the Tempest Foundation, which was forwarded to them by the university president (Figure 2.7).

"I know why President Wilson is interested in this grant," said Anne, the V.P. for Physical Facilities. "He is concerned about the university's bottom line. Energy costs are eating up a major part of the university budget, and it's only going to get worse. So, the president is trying to put energy conservation projects into the pipeline, which will pay off later."

George, a professor of environmental engineering, said, "Listen, I'm as excited about this project as anyone, but we're talking about a whole change to the way we do things on this campus. We won't be able to just set up a wind generator or a few solar panels and call it a success."

"Sure. Of course not," said Anne, "I think we all know that there's no quick-fix solution to this problem. I'm sure we can make some smaller changes right now, like encouraging people to conserve energy on campus or take the bus, but some changes will need to be made over years, maybe decades."

Karen, a professor of social sciences, agreed, "These things do take time, but people can adapt if they buy into the concept. Fortunately, on a college campus, almost everyone will agree that we need to do something to help the environment. And, as a university, we have economic reasons for making this happen. Those are some strong motivators."

Tim, the Chair of the Student Environmental Council, nodded. "I know the students would like to do this. A project like this one offers plenty of opportunities to do something good for the university and the Earth while picking up some experience."

Calvin, the local contractor, said, "Hey, those of us with local construction companies are interested too. Of course, we're interested in doing the work, but many of us also want to stay on top of these issues about sustainable energy."

The members of the team seemed to all understand that the problem they were trying to solve was complex. Fortunately, the writing of the grant proposal would help them put their ideas on paper and focus their efforts.

Figure 2.7: The Request for Proposals (RFP) from the Tempest Foundation

The Tempest Foundation

Sustainable Development and Conservation: Guidelines for Grants

Following a strategic review of the Foundation's earlier support for conservation issues, the Tempest Foundation has decided to focus its grantmaking on issues involving the climate crisis caused by humans. We believe strongly that the climate crisis is the greatest threat to the planet's ecosystems and human survival. The core of the Foundation's grantmaking will be devoted to support for research and projects that will promote long-term environmental sustainability.

In the past, we have supported a wide range of conservation projects. We will now be focusing our grantmaking on projects that have the greatest potential for lasting impact on this planet. Therefore, the Tempest Foundation will concentrate its grants on projects that promise real change. We do not need additional studies to reaffirm that climate change is a dire threat to the planet's ecosystems. We consider the facts of climate change to be settled science. Instead, we want to fund research and projects that will take purposeful steps toward solving the problem of human-caused climate change. Fortunately, recent advances in wind power, tidal power, solar energy, nuclear energy, and emergent technologies are making non-carbon forms of energy cheaper and more usable than older carbon-based energy forms.

The most attractive projects to the Foundation will be ones that inspire other projects. Our intent is to use our funding to generate other initiatives that go beyond the initial funding. We are especially interested in projects that can be transferred and repeated elsewhere.

Annual support for individual projects typically ranges from $100,000 to $1.6 million. The Tempest Foundation is especially attracted to projects that can attract funding from other sources.

Application Process

The Foundation meets to consider grant proposals four times a year. As described in the section "How to Apply for Grants" (http://itempestfoundation.org/howtoapplyforagrant), the Foundation will only consider brief proposals with narratives under fifteen pages. After the review, the board of directors will decide a) whether funding can be extended based on the proposal or b) whether a more detailed proposal will be requested. Grants with narratives of more than fifteen pages will be considered by invitation only.

Questions about the Foundation and its grantmaking can be directed to John Philips, Tempest Foundation Administrator, at forinfo@itempestfoundation.org. Inquiries by phone can be made at 312-555-0128.

George read the RFP from the Tempest Foundation out loud to the group. Then, they began discussing the Five-W and How questions. On his laptop, George typed their answers to the who, what, where, and when questions, putting question marks in places where they were unsure about the answers.

Afterward, they turned to the why question. Anne asked, "Why do you think the Tempest Foundation decided to offer this funding? What's changed recently that created this opportunity?"

Karen spoke up, "Well, I did a little research into the Tempest Foundation. They've always been interested in conservation issues. But recently, the foundation's board of Directors decided that the climate crisis is the single most important threat to life on this planet. So, they shifted most of their funding into this one issue."

George frowned. "OK, but why would they give us the money rather than someone else? We just want to make changes to our own campus. We're not proposing high-level research or a nationwide change here."

"That might be the key to persuading them," replied Tim. "We can show them that our project at Durango University could be used as a model for transforming other college campuses. That might make funding our project more attractive to the foundation."

The others agreed. Anne said, "That seems to be a good way to address the why question. I'll call the Point of Contact to see if our project would work for them. For now, let's work on defining our problem here at Durango University."

Using the four stasis questions, they began to isolate and identify the problem they faced. They started by tackling the first stasis question, "Is there a problem?"

Karen answered, "Of course. Energy is a worldwide problem that will be the defining challenge of the twenty-first century. Energy issues affect just about everything from climate change to political stability."

"Yes, but that's not a problem we can solve," said George. "Our problem is that our campus is completely reliant on nonrenewable energy sources, like natural gas and petroleum."

Anne said, "OK, we agree there's a problem. Let's answer the second stasis question, "What exactly is the problem?"

Tim spoke up, "Well, people need to conserve and make a conscious effort to reduce their use of energy. People don't recycle enough, and they don't take public transportation even when it's easier than driving a car."

"Tim, I don't mean to be skeptical here," Karen said, "but most people on this campus won't change their ways, even if they support the project. They might try recycling and public transportation, but most will eventually revert to their wasteful ways. People do what is most convenient. They don't think about how much energy they are using."

Calvin jumped in. "Maybe we need to do something that will get people to conserve energy whether they want to take part or not. For example, if we put solar panels on all the buildings, we could generate a significant amount of electricity without changing the way people do things around here."

George said, "But that's not going to solve our problem with the university's fleet of gas-guzzling vehicles. The university can't function without those trucks."

"Of course not," Karen responded, "but again, we can make strategic changes, perhaps over several years, that would eventually wean the fleet off gasoline. Each time a truck needs to be replaced, it could be replaced by something that uses electricity. Fuel cell vehicles might be available in the next couple of decades. We can do this gradually."

Anne looked up from her notes. "I think I'm beginning to understand our real problem. Our problem is that this campus is reliant on nonrenewable energy sources, and we don't have a realistic long-term strategy that will gradually allow us to convert to 'sustainable' uses of energy. We can ask people to conserve until we're blue in the face, but they will only do so much. Instead, the university needs to see sustainability as a strategic goal that will be carried out over many years. Of course, we can ask people to make changes to their lifestyles right now, like using less water, recycling, and taking public transportation. We can use incentives to encourage them not to drive their cars to campus. But, in the end, the university needs to commit itself to making strategic infrastructure changes over a longer time period."

The others agreed with Anne's definition of the problem. The problem wasn't that people on campus were making the wrong decisions about their energy usage and their use of other resources. The problem was that the campus infrastructure was dependent on nonrenewable sources of energy. The campus was not designed in a way that made the sustainable use of energy possible.

Feeling like they accomplished something by isolating the problem, the task force decided to tackle the third stasis question, "What kind of problem is it?"

George started, "I don't know about you folks, but I still don't feel like I have a good grasp of the problem at this point. But, like all of you, I don't want to just spin our wheels waiting for a comprehensive solution. I feel like we need to make some short-term changes now and set longer-term changes in motion."

"I agree," said Tim. "I think we're talking about creating a strategic plan of some kind. So, we're looking at writing a planning proposal."

"Is that what the grant would pay for?" Karen asked.

"Maybe," said Tim. "Our grant proposal would sketch out a general description of the full plan. We would ask for money to do more research and develop a strategic plan."

"That makes sense," said George. "We're all smart people, but there's no way the five of us are going to come up with that kind of comprehensive plan. We need to get more people involved. The grant could pay for us to research the campus's current energy use. Then, it might pay for the university to put together the strategic plan."

Calvin was looking a little disappointed. "I was hoping the grant money would at least buy a few solar panels or something."

Anne looked over at him. "I feel the same way," she said, "but I think Tim and George are right. Our problem is that we don't have a strategic plan in place. Creating that kind of plan is the only way we are going to make lasting, long-term changes to this campus. If the grant money allows us to do some research and develop a strategic plan for a carbon-neutral campus, then the university can begin using that plan to make strategic changes to the campus. Short-term changes won't get us far, but a long-term smarter plan could guide the transformation of this campus.

They decided to write a planning proposal. Their grant proposal to the Tempest Foundation would first describe their research methods for gaining insight into the problem. Then, the proposal would sketch out the boundaries of a strategic plan. They would use the grant to pay for developing a full strategic plan.

By the end of the meeting, they felt like they were making some headway. They agreed to meet again the next Friday to work on the proposal.

3 GETTING STARTED ON YOUR PROPOSAL

Topic, Angle, Purpose, Readers, and Context

People new to writing proposals and grants often want to dive in and start putting words and sentences on the screen. The problem with this approach is that they haven't fully figured out what they are writing about, who they are writing to, and what they want to achieve. Experienced proposal and grant writers will start by doing preliminary research and thinking about the rhetorical situation the proposal or grant is responding to.

The rhetorical situation for a proposal or grant involves five elements:

Topic: What is my proposal about? What is it not about?

Angle: What is new about this issue or makes it especially important right now?

Purpose: What primary objective is this proposal trying to achieve?

Readers: Who will be the readers of this proposal?

Context: Where and how will these readers use my proposal, and how will that context shape how they will read it?

When preparing to write a proposal, you and your team should answer these questions to fully understand the situations in which the proposal will be used. You can also use artificial intelligence (AI) applications with integrated Internet search engines to help you generate possible answers to these questions, allowing you to think more broadly about how to understand your proposal's topic, any potential angles, the purpose, the target readers, and the likely contexts in which they will read and use the proposal.

Understanding the rhetorical situation will help you determine your readers' wants and needs. Your readers are the people who can say yes to your proposal, so you need to see things from their perspective. By exploring the topic, angle, purpose, readers, and context of use, you can develop a proposal that will get them excited about your plan to solve their problem or take advantage of a new opportunity.

In this chapter, you will learn how to use rhetorical invention strategies and do research to answer the five rhetorical situation questions (topic, angle, purpose, readers, context of use). This will allow you to fill out the Getting Started worksheet shown in Figure 3.1 to get your team on the same page.

Figure 3.1: Getting Started Worksheet

Getting Started

Topic
What exactly is my proposal about? What is my proposal not about?

Angle
What is new or has changed about this topic? What has happened recently that makes it interesting right now?

Purpose
In one sentence, what is the purpose of my proposal? What is the main point I am trying to demonstrate or prove? Finish the following sentence:

The purpose of this proposal is to…

Readers
Who are the primary readers of this proposal? What are their needs, values, and attitudes toward my topic?

Context of Use
How does the place (where and when) affect how they will read my proposal? How does the medium shape how they will read? What economic or social trends will shape how they interpret what I have to say?

Topic: What Is Need to Know vs. Want to Tell Information?

Essentially, the *topic* is what your proposal or grant is about (and not about). In Chapter 2, you learned how to determine the stasis of a problem or opportunity. Answers to the four stasis questions (i.e., Is there a problem? What exactly is the problem? How serious is the problem? and "What type of proposal is needed?") will provide insight into your proposal's topic.

Now that you have a good idea of the problem you are trying to solve, you should start thinking about the boundaries or scope of that problem. Specifically, ask yourself two questions:

- What information do my readers *need to know* to say yes to this proposal?
- What information about this topic, no matter how interesting, is *not* needed to make a decision about this proposal?

These questions are important because readers evaluate proposals from a need-to-know perspective. In other words, they only want to spend time processing information that will help them make an informed decision. Writers, on the other hand, sometimes approach a text from a want-to-tell point of view. After spending weeks, perhaps even months, researching and collecting information, writers are eager to tell the readers about everything they have found. Unfortunately, all that extra want-to-tell information will get in the way of the readers' desire to find the need-to-know information that will allow them to make an informed decision.

You might find it helpful to use a brainstorming list to separate the need-to-know from the want-to-tell information. Here's how to make this kind of list:

1. Open a blank document in your word processing software or on Google Docs, MS Teams, Discord, or Slack if you're working in a team.

2. Make a list of everything you and your team can think of related to your proposal's topic.

3. Bold all the items you believe the readers *must* know to say yes to the proposal.

4. Move any non-bold items to the bottom of your list.

You can also use an AI application to help you with this kind of brainstorming. Ask the AI to make a list of, say, one hundred items or issues related to the topic of your proposal. The AI algorithm will likely suggest things you hadn't considered or didn't realize were related to your topic.

Angle: What Is New or Has Changed Recently?

Your *angle* is something new or that has changed recently about your topic. If you are responding to a Request for Proposals (RFP), ask yourself what is new or has changed recently that prompted your potential customer or client to request proposals. Maybe a new technology has emerged. Maybe a government policy changed. Maybe a shift in the market happened. What changed?

You can run Internet searches and use an AI application to explore possible angles. Ask a question like, "What has changed recently related to X?" where X is the topic of your proposal. Then you can ask, "Why did this change happen to X?" The search engine and AI application will probably return five to ten reasons why something is new or has recently changed about your topic.

At this point, you may see a few potential angles that your proposal could use. You don't need to settle on one angle right now, so write down your two or three best ideas. As you research and think about your topic, one of these angles will probably reveal itself as the best one.

Why do you need an angle? A clear angle will help show the readers that you understand why they are looking for someone to provide them with a specific service or product. Likewise, if you are writing a grant proposal, your angle will show the reviewers that you're aware of the significant changes happening that warrant new research or a project to fill an emerging knowledge gap or social need.

Purpose: What Is the Project's Primary Objective?

Your purpose is what you, your customer, or your clients want the project to achieve. Your purpose statement can also be called your primary objective. Here are a few examples:

> The purpose of this proposal is to offer a plan for developing a nonhazardous foam that cleans the fiberglass blades on wind turbines.

> Our aim is to secure funding from the National Institutes of Health to study the effects of depression on teenagers whose parents are regular users of prescription opioids.

> The primary objective of the Hack Your Life Summer Camp will be to introduce students from low-income communities to careers in computer programming and engineering.

One of the most common reasons reviewers reject proposals is that the writers were not completely clear about the project's purpose. Therefore, you want to be explicit about what the proposed project is designed to achieve.

The secret to writing a good purpose statement is stating your purpose in one sentence: "The purpose of this proposal is to. . . ." If you cannot squeeze your purpose into one sentence, your proposal might not be focused enough for the readers.

Fortunately, once you have hammered down the purpose of your project into a one-sentence statement, you will have established a foundation for the entire proposal. As you write, you can look back at your statement of purpose to see if the project achieves what you set out to do.

Reader Analysis: Who Can Say Yes to the Proposal?

Professional proposal and grant writers will tell you that developing a complete understanding of your target readers is the most important part of the proposal-writing process. When analyzing your readers, you should first recognize that different levels of readers will read and use your proposal. They can be sorted into four categories that reflect their importance in the decision-making process.

Primary Readers (Decision Makers)

Your primary readers are the people who will ultimately say yes (or no) to your proposal. Usually, your proposal's primary readers are one or two people or perhaps a committee. They are the decision-makers because they are most responsible for determining whether your project will be accepted or funded. If you are unsure who your primary readers are, ask yourself who has the power to accept your proposal. Bottom line: Who can say yes?

Secondary Readers (Advisors)

Secondary readers tend to be experts who can influence the primary readers to accept or reject your proposal. Think of your secondary readers as trusted advisors to whom your primary readers might turn for guidance. They could be engineers, subject matter experts, consultants, accountants, or lawyers who check over the methods, facts, figures, and budget of the proposal or grant.

Your proposal's secondary readers often have their own priorities. As experts, they are usually looking for more specialized information than the primary readers. For example, the senior engineers at a company could significantly influence whether a proposal is accepted because they know better than the primary readers whether your approach or technology will work. As you write the proposal, you should keep those senior engineers in mind, even though they may not be the people who can ultimately say yes to your ideas.

Tertiary Readers (Evaluators)

Tertiary readers are the people who you may not expect to read your proposal but might have a stake in what you are proposing. Tertiary readers could include journalists, outside lawyers, program assessors, historians, politicians, local residents, or your company's competitors. At first, keeping the interests of these distant readers in mind might seem a bit odd, but tertiary readers can sabotage (or support) your plans. You should always identify these potential readers to ensure that you are not exposing yourself, your company, or your organization to external challenges, bad publicity, or lawsuits.

Gatekeepers (Supervisors)

Gatekeepers are the readers who have the most direct influence over you personally and your team. They include your direct supervisor, your company's accountant, your company's legal counsel, and maybe even your company's CEO. They might include the board of directors of your organization. Gatekeepers need to sign off on your proposal before it is sent to the primary readers.

Again, it might seem strange to consider the needs of these readers as you invent your ideas. But if your proposal or grant is unacceptable to your accounting team or the legal department, then the primary readers will never have the chance to say yes. Before you start writing your proposal or grant, you need to figure out what these gatekeepers want to see in the proposal.

Otherwise, the proposal may get caught in an endless loop of revisions as gatekeepers ask for further changes.

These four types of readers represent various individuals who may influence how you craft your proposal. How can you identify all these different people and prioritize their interests? One way is to use a Writer-Centered Worksheet to help you identify and sort out all these different readers (Figure 3.2).[1]

Here is how the worksheet is used:

1. Place yourself and your organization in the half-circle labeled *Writer*.

2. In the arch labeled *Primary*, write down the primary readers, preferably by name, who will be directly responsible for deciding to accept or reject your proposal. You should name only a few primary readers because not many will have the power to say yes or no to your proposal.

3. In the arch labeled *Secondary*, write down all the readers who might serve as advisors to the primary readers. Think about the people the primary readers might turn to for information or advice. In most cases, you will find that there are many more secondary readers than primary readers.

4. In the arch labeled *Tertiary*, try to imagine anyone else, no matter how remote, who might have an interest or stake in your proposal, even if you never intended for them to get a copy. These tertiary readers include your company's competitors, journalists, and potentially hostile lawyers.

5. In the *Gatekeepers* box, list your supervisors and others who must approve your proposal at your company or organization before it is sent to the primary readers.

AI applications can help you identify possible readers in each of these arches. You can ask the AI, "Who would likely make a decision about X in the Y corporation or organization?" or "What type of specialist would serve as an advisor to the CEO in X corporation? or "Who in the local area might be concerned about or interested in a project that does X?" Based on information publicly available, an AI application with an integrated search engine should be able to find information about the people to whom you are writing.

The Writer-Centered Worksheet in Figure 3.2 will help you visualize your potential readers by spatially showing their relationship to you and your team. Your primary readers are usually the most important part of your audience, so they occupy the circle closest to you. The secondary and tertiary readers occupy places a bit further away. The gatekeepers are off to the side because they are not the intended readers of your proposal, but you will need their approval to move the project forward.

---

1. These worksheets are based on diagrams that originally appeared in Mathes and Stevenson's classic textbook, *Designing Technical Reports*, 2e (1991) p. 38.

Figure 3.2: Writer-Centered Audience Analysis Worksheet

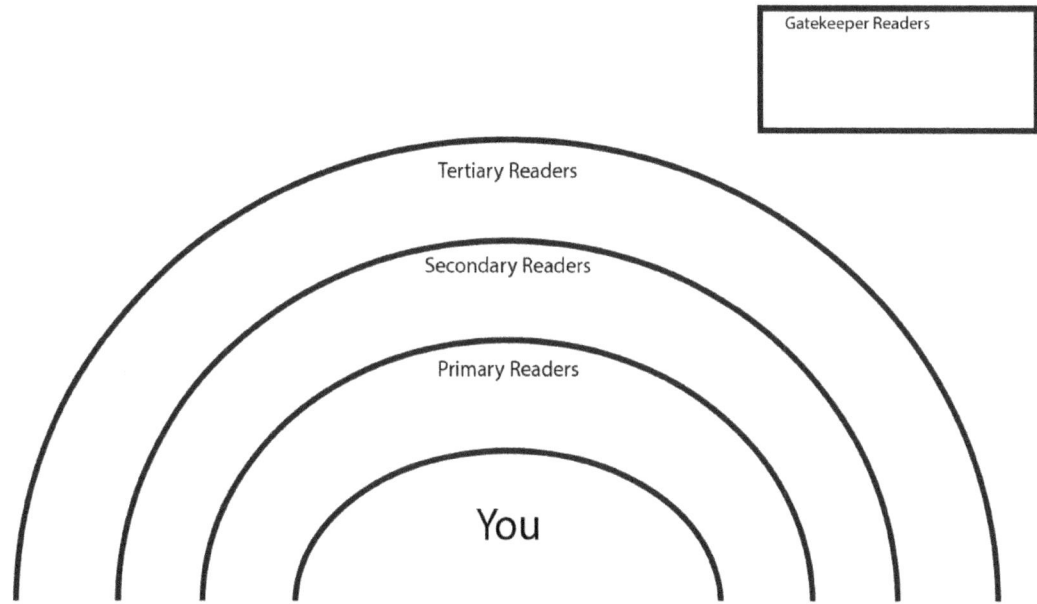

The process is the same for grant proposals. You may not know the names of your primary readers, but you can make some educated guesses about the types of people who will be reading the proposal. Put your best guesses in the Writer-Centered Worksheet. Remember that many funding sources publish the names and backgrounds of officers and reviewers on their websites. Based on this information, you can guess who is likely to read the proposal, and you can use AI and the Internet to find more information about them.

Now, it's time to consider the psychology of the people you identified in the writer-centered worksheet. You should try to get inside their minds and figure out why they might say yes or no to your proposal. To begin, keep in mind that readers of proposals and grants react positively or negatively on four levels: motives, values, attitudes, and emotions.

Motives

Readers are motivated to say yes when they think a project will improve their personal, professional, or organizational lives. For example, perhaps a particular reader is motivated by a higher profit margin. A successful business proposal would address that motivation by stressing the enhanced profit margin created by the proposed project. Another reader might be motivated to fight poverty, so a successful grant proposal written to this reader might show how the project will lift people out of poverty. Here, we use the word *motive* to suggest that people have reasons for doing things. When you identify someone's motives, you will know what moves them to act.

Values

Readers react positively (or negatively) when an idea or project might impact their personal, professional, or organizational values. So, you should spend some time identifying what your readers value. Usually, it is not that difficult to figure out your readers' values. The websites and social media feeds for companies and funding sources usually include policy statements, mission statements, and ethics policies. These documents publicly spell out the values that the organization holds. Likewise, an individual reader's professional or personal values can often be found in their corporate bios, public statements, speeches, or past actions.

Attitudes

Readers typically start with a positive or negative attitude toward a given proposal. Sometimes, they look forward to moving into new markets or solving a long-standing problem. However, in other cases, readers will approach a proposal with a negative attitude. For example, the in-house experts at a client company might have a negative attitude toward your proposal because they think their CEO's decision to solicit proposals implies a lack of faith in their in-house ability to solve the problem themselves. Of course, your readers' attitudes are always hard to judge, but you should pay close attention to what your readers say and how they say it. Sometimes, your readers' language or tone can tell you a significant amount about their attitude toward you, the project, and your proposal.

Emotions

Readers also react emotionally to proposals in ways that go beyond the simple logic or the costs of your proposal. For instance, if you propose renovating a landmark building, the readers may have strong positive or negative emotions about the project. They may feel joy, frustration, pride, or even anger. You should always take these emotions into account as you develop your proposal. Positive emotions can be used to energize your proposal and motivate your readers. Negative emotions can be addressed and softened by stressing the benefits of solving the problem.

Again, a worksheet can help you sort out the various readers' complex motives, values, attitudes, and emotions. By listing the different readers in a Reader Analysis Worksheet (Figure 3.3), you can anticipate the psychological factors that will affect their reactions.

Figure 3.3: Reader Analysis Worksheet

Readers	Needs (Motives)	Values	Attitudes	Emotions
Primary Readers				
Secondary Readers				
Tertiary Readers				
Gatekeepers				

Using the Reader Analysis Worksheet is simple:

1. In the left-hand column, list the readers you identified in the Writer-Centered Worksheet.

2. Working from left to right, fill in what you know about each reader's motives, values, attitudes, and emotions.

3. If you can't fill in one of the boxes, just put a question mark in that space. Question marks signal places where you need to do more research about your readers.

Once you fill out the worksheet, you should have a good idea about some themes and emotions you can use in your proposal to encourage the readers to say yes to your ideas.

Context Analysis: What External Factors Will Influence the Readers?

After completing a reader analysis, you should consider the external contextual factors that may affect your proposal and its. Your context analysis is related to your reader analysis because the context involves the physical, economic, ethical, and political environments in which the readers will evaluate your proposal or grant. These contextual factors will shape how they interpret and react to your ideas. A Context Analysis Worksheet like the one shown in Figure 3.4 can help you find these external factors.

Physical Context

Professional proposal and grant writers will often try to visualize the physical contexts in which their document might be used. Will the readers access the proposal on paper, computer screens, or their mobile phones? Will they read and discuss the proposal in a large meeting room, at their desks, or in a small conference room? Are the readers expecting a longer document with all the details or a shorter document with detailed appendices? These elements of the physical context will influence how you organize and design your proposal.

Economic Context

Of course, the bottom line is the bottom line for any proposal. Your readers will always have costs and other economic issues in mind as they consider your proposal. You should carefully consider the economic factors influencing your client or the funding source. Sometimes, an expensive project might solve all their problems, but your customer, client, or funding source can only agree to something more affordable.

On a larger scale, consider the economic factors impacting your customer's or client's industry. Do some research to fully understand trends in the current market, especially the impacts of those trends on the client's products or services. If you are writing a grant proposal to a foundation, you may be able to find out how much money the funding source has given for similar projects in the past.

Overall, you should always pay close attention to the economic issues in a proposal. Your readers will certainly have these issues front and center in their minds.

Ethical Context

Clients will tend to shy away from proposals that sound ethically questionable. In an increasingly litigious society, the ethics of any project are critically important. Therefore, you need to be mindful of plans that might cause ethical issues, harm your customers' or clients' image, or leave them open to liability lawsuits. As you analyze the context, try to identify any potential ethical problems, even the most obscure.

Political Context

In proposals, political issues come into play on two levels. First, as corporate citizens, most company executives and boards of directors know the national, local, and industrial political issues that affect their business or organization. Proposal writers should be well aware of the politics in a particular industry and how they play out at the local, state, and federal levels. Grant writers should consider how the project they are proposing will affect the political status quo.

The second level of politics involves office politics, which shape how proposals are reviewed. You should always remember that any proposal usually treads on someone's turf. Sometimes, personal networks or existing relationships might give one proposal an edge over others. Your competitors, of course, will be trying to outflank you by using their connections. These local political issues are unavoidable, but you should be aware of them to better shape what you are offering and your persuasion strategies.

To help you sort out all these external factors, the Context Analysis Worksheet in Figure 3.4 is divided into three levels, primary readers, industry/community, and writer's organization, to represent the various levels on which these contextual issues tend to influence the readers.

Put question marks in spaces where you do not know enough about the readers' context. These question marks signal places where you may need to do more research on your readers.

Again, AI applications help you figure out the potential physical, economic, ethical, and political factors that might be at play with your proposal. You can ask an AI application questions like, "What ethical issues might be involved with a project that does X?" or "How might a project that changes X also put pressure on politics in the Y area?" You can use AI to help you identify contextual issues and pressure points that might not be immediately apparent to you or your team.

Figure 3.4: Context Analysis Worksheet

	Physical	Economic	Political	Ethical
Primary Readers				
Industry/ Community				
Writer's Organization				

Focusing a Writing Team

You will be well on your way to writing an effective proposal if you sit down with your co-workers and agree on the proposal's topic, angle, purpose, readers, and context. Your writing team will be much more focused and efficient if you do these five things:

Define Your Topic: Use the four stasis questions discussed in Chapter 2 to figure out (a) if there is a problem, (b) what the exact problem is, (c) how serious the problem is, and (d) what type of proposal would be appropriate. Then, discuss what information the readers need to know to make a decision on your ideas. Also, try to identify information the readers do not need to know.

Come Up with a New Angle on Your Topic: Research and brainstorm to figure out what is new or what has recently changed about your topic. Your new angle will create a fresh and forward-thinking impression in your readers' minds.

State Your Purpose: State the purpose of the proposal in one sentence, period. That is, complete the following phrase: "The purpose of this proposal is to...." If you need more than one sentence to state your purpose, your understanding of what needs to be done might not be focused enough.

Analyze Your Readers: Identify the various readers (primary, secondary, tertiary, gatekeepers) and their characteristics (motives, values, attitudes, and emotions). To identify the readers and their characteristics, fill out the Writer-Centered Worksheet (Figure 3.2) and the Reader Analysis Worksheet (Figure 3.3).

Analyze Your Reader's Context: Identify the various contextual issues (physical, economic, ethical, and political) that will influence how the readers will interpret and react to your proposal. Fill out the Contextual Analysis Worksheet to sort these issues into levels that influence the primary readers, the industry/community, and you and your organization (Figure 3.4).

Doing this kind of analysis with your team will start your proposal-writing process on the right foot.

Chapter Summary and Looking Ahead

This chapter and the previous chapter were designed to help you start thinking about the problem/opportunity your proposal or grant is pursuing. You learned how to clearly define the problem and anticipate the rhetorical situation in which your document will be used. Now, it is time to start learning how to write your proposal. The next chapter will discuss how to write your proposal's introduction.

Try This Out!

1. Find a proposal or grant proposal on the Internet or from your workplace. Write a two-page analysis in which you discuss how the writers handled the proposal's topic, angle, purpose, readers, and context. Can you find places in the proposal tailored to the specific readers or context? Do you think the proposal achieves its purpose? Are there places in the proposal where the writers stray from their purpose? How might the proposal be improved to better fit its rhetorical situation?

2. With a team, choose a problem on campus, at your workplace, or in your community that could benefit from grant funding. Analyze the rhetorical situation in which that grant proposal would need to operate. Use a Writer-Centered Worksheet and a Reader Analysis Worksheet to identify the proposal's primary and secondary readers and their characteristics. Then, fill out a Context Analysis Worksheet to work through the physical, economic, political, and ethical factors that might influence the primary and secondary readers. Write an email to your instructor summarizing the important issues related to the proposal's readers and context.

3. Fill out the Getting Started worksheet based on your understanding of the rhetorical situation for your proposal. Then, use an AI application to help you answer questions about the rhetorical situation. Where were your and the AI application's answers similar? In what ways were the answers different? Write an email to your instructor in which you discuss those similarities and differences. Did the AI application generate some information that could be useful? Did it steer you a different and perhaps wrong way?

4. Writing with a team can be challenging. How might you use the methods and worksheets in this chapter to help you start a proposal with a team of others? What are three things you would do differently when collaborating with a team rather than working on your own?

5. Study the RFP in Question 6 at the end of Chapter 2. What are the contextual issues (physical, economic, political, and ethical) that might be shaping this rhetorical situation? What political issues would you need to keep in mind as you write a proposal for this RFP? What ethical issues might also be involved?

Case Study: The Carbon Neutral Campus Project—What Is the Rhetorical Situation?

In their previous meeting, the Carbon Neutral Campus grant writing team agreed that Durango University's problem was the lack of a long-term strategic plan for eliminating or offsetting the greenhouse gases produced on campus.

They agreed that most people on campus would not voluntarily change their energy consumption significantly, even if they supported the Carbon Neutral Campus concept. As a result, the university itself would need to renovate the campus infrastructure to encourage more energy conservation and the use of renewable energy sources.

"All right," said Anne, as she grabbed a dry-erase marker and went up to the whiteboard in the room, "let's start by figuring out the topic, angle, purpose, readers, and context for this proposal."

They used the Getting Started worksheet to help them fill out the rhetorical situation:

Topic: Transforming the Durango University campus from a carbon emitter into one that produces net-zero carbon emissions

Angle: Non-carbon energy sources have become more affordable than carbon-based energy due to advances in solar, wind, and battery technologies, as well as more energy-efficient cars, buses, and electronics.

Purpose: The Carbon Neutral Campus Project's purpose is to develop a strategic plan that guides the long-term conversion of the campus to renewable and sustainable energy sources.

Readers: Tempest Foundation Board of Directors

Context: The directors will read the proposal in their offices, probably on a computer screen. They will discuss the document in meetings. They might also bring a copy of the proposal when visiting the campus.

With these elements of the rhetorical situation identified, they then studied each element individually.

Topic

Defining the proposal's topic took a bit more time than they expected. They all agreed that issues involving renewable energy sources, such as solar power, wind power, and geothermal heating, would be important in the proposal. However, they had trouble deciding whether related issues like improved campus recycling programs should also be part of the plan.

Karen said, "It's important that we add a few goals that are reachable in the short term. Better recycling systems and encouraging people not to drive their cars to campus are things we can do right now."

Calvin was skeptical. "I'm just concerned that these smaller issues might distract from our larger goals. You know how these things happen. We'll see some extra recycling bins and a few signs about taking the bus. The rest might be forgotten over time."

"That's why we need to find ways to make this strategic plan an integral part of the infrastructure and mission of this university," said Anne.

George added, "Yeah, it needs to be more than just a plan. This needs to be a core objective of the university—a statement that connects with how the university does business."

"I'm reluctant to say this," said Tim, "but I think we need to narrow our topic to energy issues, cutting out non-energy issues like recycling."

"What?" said Karen. "Recycling is very important."

"I agree," said Tim, "but energy issues seem to be our main issue in this proposal. I believe we need to focus on issues that directly involve converting the campus to renewable energy sources."

The group debated whether non-energy issues like recycling should be included in the grant. They decided Tim was probably right, so they crossed out any items that didn't directly address energy-related issues.

Anne said, "Something else I think we should avoid is looking outside of campus. As much as I care about the Amazon rainforests or sustainable farming practices in Africa, we should focus on our Durango campus."

"Agreed," said George. The strategic plan should only concern issues that directly impact the Durango University campus. Other issues are important, but including them would make our proposal seem unfocused to the readers.

Angle

The proposal's angle also took more discussion than expected. At first, Tim said, "Well, the climate crisis isn't getting any better. That's an angle." The others nodded in agreement.

After a moment, Calvin spoke up. "Yeah, but that's not what's new about this topic. Even though climate change has been a serious problem for decades, people have become immune to the idea that the planet is warming. They may agree it's serious, but they think the problem is many years off or don't know what to do about it."

"That's kind of amazing," said Tim. We see the effects of the climate crisis all around us: forest fires, flooding, drought. It's one thing after another."

Karen asked, "OK. What's different that makes this topic interesting right now? Along these lines, what does the Tempest Foundation think is new about this topic, and why have they shifted all their funding to climate-related projects?"

George said, "The RFP mentions recent technological advances that allow us to do something about climate-related issues. Right now, solar energy panels are much cheaper than in the past. Wind power is growing exponentially. In the mountains, geothermal energy is right under our feet. Even nuclear power seems to be making a comeback with smaller nuclear power plants."

Calvin jumped in, "That's probably our angle. We've seen huge breakthroughs in zero-carbon energy technologies. So-called 'alternative energies' were expensive in the past, but now they're cheaper than the older dirty forms of energy. Right now, energy is just energy. There's nothing 'alternative' about these sustainable ways of generating power."

The team decided to use technological breakthroughs as a new angle to argue that the time is now to take big steps to address the climate crisis.

Purpose

They then defined the purpose of the Carbon Neutral Campus Project. What did they want the project to do? Confining themselves to one sentence, they hammered their purpose into a clear, crisp statement: "The purpose of the Carbon Neutral Campus Project is to develop a strategic plan that guides the long-term conversion of the campus to renewable and sustainable energy sources."

Though generic, this statement of purpose offered them two immediate tools for writing the proposal. First, it specified the overall purpose of the proposal itself. Second, it gave them a statement that would help them slice away any nonessential details, allowing them to focus on need-to-know information.

Readers

They then turned to the Writer-Centered Worksheet to identify their other potential readers (Figure 3.5). The primary readers would be the Tempest Foundation board of directors, who would decide whether to fund the grant or not. These readers at the Tempest Foundation were the decision-makers who would say yes to their ideas and fund their projects.

The team wasn't sure who the secondary readers might be. They speculated that the Tempest Foundation would likely hire advisors or consultants to determine whether the proposed projects were feasible. These secondary readers might be consultants, engineers, architects, or urban planners specializing in environmental projects.

Figure 3.5: Writer-Centered Audience Analysis Worksheet (Completed)

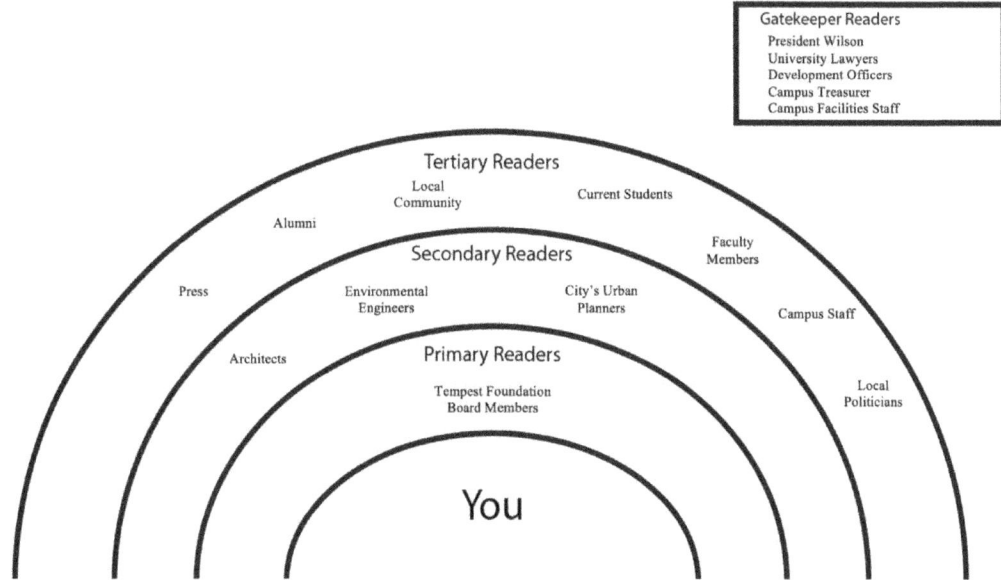

At first, the tertiary readers seemed a bit more problematic. Who else might be interested in obtaining a copy of the proposal? Karen pointed out that the media might be interested in a copy. "They will get a copy from somewhere, whether we send it to them or not. We need to keep them in mind as we're writing."

Calvin pointed out that local contractors would also be interested in a copy of the proposal because they would want to bid on any future contracts. Tim mentioned that other universities might want a copy to use as a model for writing similar grants.

George spoke up, "Hey, don't forget our alumni. The Alumni Association will be interested in this project, maybe as a way to fundraise and build the university's brand. Plus, we need to always remember that resistance from alumni can get in the way of a transformational project like this one."

Finally, they listed several gatekeepers. Anne pointed out that President Wilson was probably the most influential gatekeeper. "And, as VP for Physical Facilities, I guess I'm a gatekeeper, too." They also listed the university's accountants, legal counsel, and development officers. These experts would need to review the grant proposal before it went to the Tempest Foundation.

Having identified their many potential readers, they finished filling out a Reader Analysis Worksheet.

Context

With their readers identified and described, the team decided to use a Context Analysis Worksheet to look more closely at the situations in which the proposal would be used. It wasn't long before they realized that the context of the proposal was very complex.

The Context Analysis Worksheet highlighted the economic and political issues that would influence how the readers interpreted the proposal. The team believed the board of directors at the Tempest Foundation would be interested in a project like this one, but the board might question the university's economic and political will to follow through and implement the strategic plan.

George said, "We need to show the Tempest Foundation that everyone, from the university president to the students, is interested in transforming Durango University into a carbon-neutral campus."

Karen added, "We also should mention that we aren't expecting the Tempest Foundation to pay the whole bill. The funding they give us would help us develop the strategic plan, but the university would be responsible for raising the money to implement these changes."

The physical and ethical issues seemed manageable. Physically, the proposal would be used in offices, planning meetings, and perhaps on-campus visits. A copy of the proposal might be put on a website for easy access. As for ethical issues, the movement toward a sustainable campus seemed to be a plus from an ethical standpoint. However, they would need to be careful to hear diverse viewpoints about the proposed changes to the campus.

George said, "Perhaps we need to plan some campus meetings to solicit feedback on the near-final version of the plan. That might help us avoid the ethical pitfalls that can emerge with these kinds of well-intentioned changes to the campus."

Anne agreed. "We don't want to trample someone's rights in our eagerness to do the right thing for the environment."

Defining the rhetorical situation for their proposal took them an hour and a half. When they finished, they had developed a much richer sense of the content, purpose, and social/political factors surrounding their proposal. The foundation for writing the proposal had been set.

4 WRITING THE INTRODUCTION

Telling Them What You're Going to Tell Them

As the speech writers' saying goes, "You need to tell them what you're going to tell them. Tell them. Then, tell them what you told them." In simple terms, a proposal's introduction tells them what you're going to tell them.

At some point in your life, you've probably been told, "Never write the introduction first!" This advice supposes you won't know what you will say in the introduction until you have written a rough draft of your document's body.

This advice is well-meaning, but it causes some problems. You see, the introduction is where you will state your proposal's topic, purpose, and main point, among a few other important things. So, jumping straight into drafting the body means you may not yet have those important things worked out for yourself.

There's a better way, especially when you are writing a proposal. An introduction for a proposal will usually make up to seven moves:

Move 1: Grab your readers' attention.

Move 2: Define the topic of your proposal.

Move 3: Provide some background information about the topic.

Move 4: Stress the importance of the topic.

Move 5: State the purpose (primary objective) of your proposal.

Move 6: State your proposal's main point.

Move 7: Forecast the organization of the proposal.

These seven moves can be made in almost any order in an introduction, and not all are necessary. A good introduction will at least state your proposal's topic and purpose—that's the bare minimum. The other five moves will strengthen your introduction.

In this chapter, you will begin drafting your proposal by writing out these seven moves. These moves will help you "tell them what you're going to tell them," laying the foundation for the body of your proposal. After reading your introduction, the readers should know what your proposal aims to achieve and why.

You should never underestimate the importance of the introduction because it often makes or breaks a proposal. Your readers gain a first impression based on your introduction. If you make a positive first impression, chances are good your readers will read "with the grain," looking for evidence that your proposal is worth considering. A bad first impression, though, means they will be reading "against the grain," which means they may start looking for reasons to reject your proposal.

Move 1: Grab Your Readers' Attention

Most proposal writers like to start with a grabber (also known as a *hook* or a *lede*) because they want to grab the readers' attention. A grabber typically asks an intriguing question or says something that gets the readers' attention. Some common grabbers used in proposals include the following:

- **Ask an intriguing question**—"If you knew you could double your return on investment in three years, would you take that chance?"
- **State a startling statistic**—"In a recent online survey, 28.3% of GenMark employees reported being sexually harassed at work within the past three years."
- **Make a compelling statement**—"Despite the common belief that homelessness is caused by drug use, only one in four people who are homeless abuse drugs (SAMHSA, 2021)."
- **Begin with a quote**—"Steve Jobs, the founder of Apple, once said, 'Your time is limited, so don't waste it living someone else's life. Don't be trapped by dogma, which is living with the results of other people's thinking. Don't let the noise of others' opinions drown out your own inner voice. And most importantly, have the courage to follow your heart and intuition.'"
- **Address the readers as "you"**—"If you are like most people, you have probably received a medical bill that seemed unfair."
- **Use a one-word cartridge grabber**—"Safety. Too many Americans don't feel they are physically and economically safe."

Your grabber should capture the readers' attention and will likely name the proposal's topic.

Move 2: Define the Topic of Your Proposal

Define your topic for the readers. For example,

> High blood pressure (HBP), also known as hypertension, occurs when the force of blood flowing through a person's arteries is consistently too high. HBP is called a "silent killer" because its victims often feel no symptoms. If left untreated, this disease can cause serious health problems, including stroke, heart attack, and damage to vital organs. People who are homeless are especially at risk because they cannot afford blood pressure medicines that are routinely paid for by health care insurance.

You might also limit the scope of your topic by saying what your topic is not about:

> In this proposal, we will not discuss how HBP impacts the general public, so we can concentrate on addressing the unique challenges faced by people who are homeless.

Highlighting what the proposal is not about will allow you to set it apart from topics that might be similar but outside the scope of the proposal. In the example above, the proposal writer wants to clarify that people who are homeless experience high blood pressure challenges that go beyond those of the general public.

Move 3: Provide Background Information about the Topic

Including a sentence or paragraph of background information is a good way to educate your readers about your topic and help them settle into reading the proposal. In a proposal's introduction, background information is typically not meant to be controversial or provocative. Instead, you can tell the readers things they already know or will generally accept without challenge.

> In a 2020 study, researchers at the University of Central Florida found that 73% of people who had been homeless for more than one year were suffering from HBP. As a result, a manageable health condition that is inexpensive to treat was causing unnecessary strain on hospital trauma centers.

Other forms of background information are available. Perhaps you might talk about the current state of the industry. Or, you might remind the readers of a previous conversation, "At our meeting on January 30, we discussed some of your options for improving..." The most effective background information simply tells the readers something they already know or will readily accept. Background information offers readers a comfortable reference point from which to start considering new or different ideas.

In most proposals, you want to provide just enough background information to make the readers comfortable with the topic. Any detailed or potentially controversial information should be saved for the body of the proposal where you can go into more depth.

Move 4: Stress the Importance of Your Proposal's Topic to the Readers

In the introduction, proposals need to capture the readers' full attention. Specifically, you want to tell the readers why the topic is important to them.

The importance of the topic can be expressed positively or negatively. The positive approach stresses the opportunity available to the readers (e.g., "This new state-of-the-art facility will allow us to significantly increase our market share."). The positive approach shows them the advantages of taking action at this particular moment.

The negative approach puts the readers on alert (e.g., "If we don't build a new facility now, we will likely lose significant market share."). Of course, alarming the readers will stress the importance of the topic, but you risk making the readers defensive or resistant to change.

Whether you use a positive or negative approach depends on who you are writing to. Most readers will respond more favorably to a positive message. However, in some cases, a negative approach may be needed, especially if the readers are aware of the problem but have been avoiding doing something about it.

Move 5: State the Purpose (Primary Objective) of Your Proposal

In Chapter 3, you learned how to write a statement of purpose on your Getting Started worksheet. That sentence or a version of it should work in your proposal's introduction. Your purpose statement tells the readers what your proposed project is designed to accomplish. That's your primary objective. In one sentence, complete the phrase, "The purpose of this proposal is to…" Then, tell them what your project will accomplish if approved and funded. For example, these again are the purpose statements used as examples in Chapter 3:

> The purpose of this proposal is to offer a plan for developing a nonhazardous foam that cleans the fiberglass or carbon-fiber reinforced blades of wind turbines.

> Our aim is to secure funding from the National Institutes of Health to study the effects of depression on teenagers whose parents are regular users of prescription opioids.

> The primary objective of the Hack Your Life Summer Camp will be to introduce students from low-income communities to careers in computer programming and engineering.

The final version of your proposal's purpose statement may or may not include a phrase like "The purpose of this proposal is to. . . ."

Move 6: State Your Proposal's Main Point

Your proposal's main point is the one idea you would like the readers to take away from reading the document. In most proposals, you should tell the readers your solution to the problem or how you will help them take advantage of an opportunity. In a research proposal, you might phrase your main point as a research question or hypothesis.

By stating your main point up front, you provide the readers with a primary claim or question that will focus their attention as they read the proposal. You are essentially saying, "Here's what we want to achieve." The rest of the proposal will lead the readers through the thought process that brought you to that main point.

Move 7: Forecast the Organization of the Proposal

Forecasting identifies the major sections in the proposal. By offering an overview of the proposal's content in the introduction, forecasting helps the readers build a cognitive framework to understand the document's structure. A forecasting statement usually looks something like this:

> In this proposal, we will first explain the causes and effects of the problem. Second, we will offer a project plan designed to solve the problem. Third, we will introduce you to

our team and describe our qualifications to handle this project. Fourth, we will conclude by explaining the costs and benefits of the project. An itemized budget appears in Appendix A.

Working Out the Seven Moves on a Worksheet

While inventing the content for the introduction, you might find it helpful to write down a sentence or two for each of these moves. The worksheet in Figure 4.1 shows how you can phrase these moves as questions. Once you have written down answers to these questions, you should be ready to write the introduction.

At this point, professional proposal writers will often stop and move on to the body. The seven moves tell them everything they need to know, for now, about what will go into the introduction. They know they still need to work out the proposal's content, organization, and style. So, right now, they won't spend a lot of time revising and wordsmithing the introduction. When they finish drafting the body and conclusion, they can return to their original seven moves with a better sense of what the rest of the proposal says.

Drafting Your Proposal's Introduction

The seven introductory moves can be made in almost any order, and you do not need to include the moves. As you are writing a rough draft of your introduction, you might go through the seven moves in the order shown in this chapter. These moves build on each other, helping you figure out what your introduction and the rest of the proposal will include. Then, when you have finished drafting the body, you can rearrange the moves to suit the story you are trying to tell in the proposal.

Putting the Moves in Order

Usually, proposals start with a grabber and some background information. But there are other ways to lead off a proposal. For example, your proposal's first sentence might boldly state your purpose up front: "In this proposal, our objective is to demonstrate that . . ." Or, perhaps you might start by identifying your topic and stressing its importance: "Children who do not get enough food to eat tend to suffer more from chronic illnesses like asthma, and they can struggle in school because of anxiety, hyperactivity, and aggression."

You can also combine moves in sentences. For example, a grabber can also be used to define the topic of your proposal. Similarly, background information can be used to stress the importance of the topic to the readers. In some situations, your purpose statement and main point can be made in the same sentence: "Our proposal introduces an innovative, eco-friendly packaging solution that significantly reduces waste and enhances brand reputation among environmentally conscious consumers."

Figure 4.1: The Seven Moves Worksheet

What are a few ways to grab the readers' attention at the beginning of this proposal?

What is the topic of this proposal? What is *not* the topic of this proposal?

What background information should the readers know about this topic?

Why is this subject important to the readers?

What is the purpose of this proposal, stated in only one sentence?

What main point is this proposal trying to make to the readers?

How will the body of the proposal be organized?

Ultimately, identifying your topic and stating your purpose are the most important moves in the introduction. Before reading the body of your proposal, your readers will need a clear idea of what you are writing about (topic) and trying to achieve (purpose). For this reason, the sentences that make these key moves should appear early or toward the end of your introduction, where the readers are paying the most attention.

If you are unsure whether your topic and purpose are clear, you should make them obvious to the readers. Use a straightforward sentence like, "In this proposal, we discuss high blood pressure and its effects on people who are homeless." And, if you want to make your purpose absolutely clear, say something like, "The purpose of this proposal is to seek funding for pop-up clinics in Albuquerque that offer access to treatment and medications to people who are homeless."

It is better to be too blunt than too subtle about the topic and purpose of your proposal.

Revising the Introduction to Make it Clearer and More Concise

Your introduction should be about half a page to a full page long—no longer. When readers begin reviewing a proposal, a mental timer starts in their heads. They begin asking questions like, "What is this proposal about? Why are these people writing to me? What is the main point? Why should I care?" The reader will lose focus if you take too long to answer these questions. Up front, they need you to tell them what you're going to tell them, so they understand what you're trying to do in the proposal.

An effective introduction should only take a couple of minutes to read. After all, if you can't answer the readers' starting questions within two minutes, you probably won't be able to hold their attention for the entire proposal. Simply state the first six moves in whatever order makes the most sense to you. Then use forecasting to transition the readers into the body of the proposal.

When revising your proposal's introduction, challenge each sentence to decide whether the readers are getting the need-to-know information they need but not too much. Don't put too much detailed information in the introduction because all those details will make it difficult for the readers to find the proposal's topic and purpose. If a sentence includes information that the readers do not need to know as they begin reading the proposal, you should move that information into the body of the proposal where it will be more useful.

Chapter Summary and Looking Ahead

In this chapter, you learned how to generate the content and draft a rough version of a proposal's introduction. The purpose of an introduction is to "tell the readers what you're going to tell them." Your proposal's introduction should create a framework that tells the readers the topic of your proposal, why it's important to them, and what you're trying to do. In the next chapter, we will discuss writing the Background section, where you will analyze the problem you're trying to solve by explaining its causes and effects.

Try This Out!

1. Find a proposal or grant proposal on the Internet or from your workplace. Cut and paste the proposal's introduction into a separate document. Then, review the introduction and highlight where each of the seven introductory moves discussed in this chapter appears. (Keep in mind that a sentence can make two or more moves). Answer these questions:
    a. In what order do the moves appear?
    b. Are any moves missing? If so, did the author seem to have a good reason for leaving out these moves?
    c. Are there any sentences that aren't making any of the seven moves? If so, do you think these sentences could be moved into the body of the proposal or removed altogether?
2. Find the sentence in which the proposal's introduction names the topic for the first time (Hint: It should appear early in the introduction.) Soon after naming the topic, does the proposal also define it and limit its scope? If not, how might you define the topic and indicate its boundaries for the readers?
3. Find the sentences in which the introduction identifies the purpose and main point of the proposal. Do they clearly explain what the proposal is trying to achieve and the main point the readers should take away from it? If not, how could the authors make the purpose and main point clearer (blunter or more straightforward) for the readers?
4. Use the seven moves described in this proposal to generate the content for an introduction you are writing. The worksheet in Figure 4.1 will help you. Write one or two sentences for each move. Then, convert your introduction into a 1-3 paragraph introduction.

Case Study: The Carbon Neutral Campus Project— How Do We Get Started?

Karen was getting restless about the grant proposal to the Tempest Foundation. She knew answering the Five-W and How Questions and analyzing the readers were important, but she was growing concerned the team would lose momentum if they didn't begin putting some words on a screen. Plus, as a writer, she was one of those people who just needed to have her fingers on the keyboard to figure out what she wanted to say.

The team set up a conference call for Friday. Karen thought the meeting would be a good time to discuss drafting a part of the proposal. So, she decided to sketch out some ideas for the introduction.

When she was a student, she remembered being told to never, never, never write the introduction first. Over the years, though, she had realized that this kind of advice was not practical for her work life. As someone who wrote every day for her job, Karen usually began by drafting the introductions of her emails, letters, and smaller reports. Using the Seven Moves Worksheet (Figure 4.1), she made the following notes:

Grabber—a good grabber could be an anecdote or shocking statistic about the effects of the climate crisis, especially on people living in the Durango University area.

Topic—converting the Durango University campus into a carbon-neutral or even a carbon-negative site. The proposal would not take on all the global issues created by the climate crisis. It would limit its scope to the campus.

Background Information—information on the effects of the climate crisis on southwestern Colorado. Details about Durango University's environmental mission and current efforts to reduce carbon emissions.

Importance of Topic—a few scary statistics and an anecdote about the effects of the climate crisis on southwestern Colorado.

Purpose of the Proposal—to acquire funding to create a strategic plan that Durango University will use to transition into a carbon-neutral campus.

Main Point—Durango University's conversion to a carbon-neutral campus would offer a great model that other universities can follow.

Forecasting—This proposal will include a Background section that analyzes the problem, a Project Plan section that describes how the project would be completed, a Qualifications section that describes our team, and a Cost and Benefits section that reviews the project's budget and deliverables.

Karen knew she needed to give these moves some more thought in the future.

She decided to email Anne to see what she thought. When they met in person, Anne was a little leery about Karen's idea of stressing the importance of "scaring" the readers. Anne said, "We need to remember that the Tempest Foundation already knows the problem is serious. That's why they are willing to put money into these kinds of projects. We don't need to scare them."

"But we need to show them that we think the problem is a very important one," said Karen. "They need to see that we take this issue very seriously."

Karen paused for a moment and decided to look for a middle ground. "Is there a way to show the reviewers that we view the climate crisis as a significant threat while also stressing the advantages of prioritizing doing something about it? In the RFP, they signal that they are more interested in taking action than doing more research or handwringing about the future."

Anne thought about that for a moment and said, "How does this sound: 'The climate crisis feels like something each of us can't solve. At Durango University, we believe we can take a critical step in the right direction by transforming our campus into a carbon-neutral site.'"

Karen said, "Maybe that's still a bit too negative, but I like how you're balancing the scary part with taking action to solve the problem. We should emphasize solving the problem more than just describing it."

"Yes," agreed Anne. "Maybe the introduction could state a few key facts about the effects of the climate crisis on southwest Colorado but not go into too much depth." On her laptop, she brought up a summary of the most recent report from the Intergovernmental Panel on Climate Change (IPCC), released in 2022. It offered specific numbers about the effects of the climate crisis in the southwestern U.S.

"This looks great," said Karen. "These facts from the IPCC will tell the Tempest Foundation something they probably already know, but this kind of background information will give us a familiar place to begin describing the Carbon Neutral Campus Project to the reviewers."

Karen and Anne circulated their notes to the team before Friday's video conference.

At the meeting, Calvin had some additions and suggestions for improvements, but he didn't want the team to get bogged down on writing an introduction right now. He said, "Listen, you've both done some nice work here. Let's keep these notes and move forward to working on the Background section of the proposal. That way, when we have a good sense of the problem and our plan, we will be in a better place to write the proposal's introduction."

On the screen, the others were nodding and agreed to that approach. They all knew they had a lot of work ahead of them. They didn't want to spend too much time working on an introduction that would change as the Carbon Neutral Campus Project evolved.

Karen and Anne had given them a solid start by using the seven introductory moves to generate content for the proposal. Now, they were ready to work on the Background section.

To see how the introduction of the Carbon Neutral Campus proposal turned out, turn to the final version in Appendix A.

5 WRITING THE BACKGROUND SECTION

Describing the Current Situation

With your proposal's introduction sketched, you are ready to start drafting the Background section. The purpose of the Background section (also sometimes called the *Current Situation* or *Narrative*) is to define the problem or opportunity and explain its causes and effects.

Depending on your readers, this section can be short or long. If your readers have a solid understanding of the current situation, this section can be shorter because you are mostly demonstrating that you, too, have a solid understanding. If your readers seem uncertain about the problem or what's causing it, the section may be longer because you will need to show them your analysis, explaining the causes and effects in more depth.

In Chapter 1, you learned that proposals tend to follow a pattern, also known as a *genre*. As shown in Figure 5.1, the Background section usually follows the introduction. Keep in mind, though, that the proposal genre is not a formula to be followed mechanically. Rather, it's a pattern that can be altered, reorganized, and stretched to fit the needs of any situation. Your customers, clients, or funding sources may ask you to follow an organizational pattern different from the one shown here. If so, you should follow their pattern, of course. But, if they don't specify a pattern, the one shown in Figure 5.1 offers a solid way to organize the major sections of a proposal.

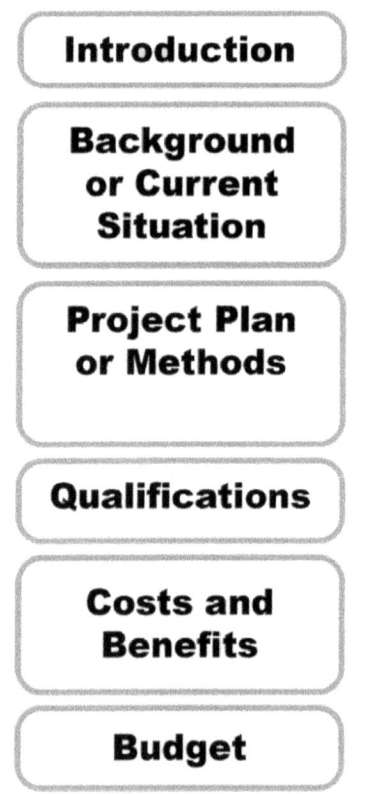

Figure 5.1: Major sections in a proposal or grant proposal.

Depending on the kind of proposal you are writing, the Background section will include different kinds of information:

- In business proposals, you need to show the customers or clients that you fully understand their current situation, thus increasing the likelihood that you will offer a reasonable plan to solve their problem or help them take advantage of an opportunity.
- In grant proposals, you may need to educate the readers about the history or background of the problem you are trying to solve, including its causes and the effects of letting it continue.

- In grant proposals for research funding, the Background section is where you will offer a literature review and describe your prior research into the topic.

Put simply, the Background section is where you will provide the readers with the background information that they need to understand the problem, its causes, and its effects. That will help you set up your project description, which you will provide in the Project Plan or Methods section.

Guidelines for Drafting the Background Section

Your main goal in the Background section is to explain to the readers the causes and effects of the problem or opportunity. As you draft the Background section, you should keep three guidelines in mind:

Guideline 1: Problems are the effects of causes.

Guideline 2: Ignored problems tend to grow worse.

Guideline 3: Blame change, not people.

Guideline 1: Problems Are the Effects of Causes

This first guideline urges you to search for the elements of change that are at work behind the current situation, specifically the causes of the problem or opportunity. Problems and opportunities do not just happen—they are almost always *caused* by something. Something in the current situation has changed to create them.

One way to identify the causes of a problem or opportunity is through a technique called *concept mapping*, or just *mapping* for short. Mapping helps you sort out the problem on a whiteboard, sheet of paper, or computer screen. That way, you can visualize the major and minor causes of the problem for yourself.

To map the current situation, go through the following steps using a whiteboard, sheet of paper, or a computer that has mapping software on it:

1. Write down the problem in the center of a board, paper, or screen and put a circle around it

2. Identify 2-5 major causes of that problem and write these causes separately around the problem. Circle them and use lines to connect them to the original problem (Figure 5.2).

3. Map out further to identify 2-5 minor causes behind each major cause. In Figure 5.2, for instance, there are four major causes (the ones closest to the problem). By mapping further, you can identify the minor causes that created each of these major causes.

4. As you map out the causes, keep asking yourself, "What changed?" As discussed in Chapter 1, proposals are tools for managing change. By paying attention to what is chang-

ing or evolving in the current situation, you will better understand how the problem or opportunity came about.

If you are using an artificial intelligence (AI) application to help you research the problem, you can ask it questions like "What are the major causes for the problem X," where X is the problem you are trying to explain. The AI will usually identify anywhere from 3-10 major causes for the problem. Then, for each of those major causes, ask the AI application to identify a few minor causes. It will likely identify 3-10 minor causes for each major cause. Add these minor causes to your map.

One word of caution: Ensure that you verify that the causes mentioned by the AI are real and that you can support them with reputable sources. Using fictitious information in your proposal will almost certainly sink its prospects. If the AI has an integrated Internet search function, you can ask it to provide you with sources for its claims.

While mapping, you can keep teasing out more minor causes for the problem. Eventually, you will find that you have developed enough detail to fully explain the problem to the readers. You shouldn't need to go beyond two or three levels of major and minor causes in your concept map.

Figure 5.2: Mapping the Causes of a Problem or Opportunity

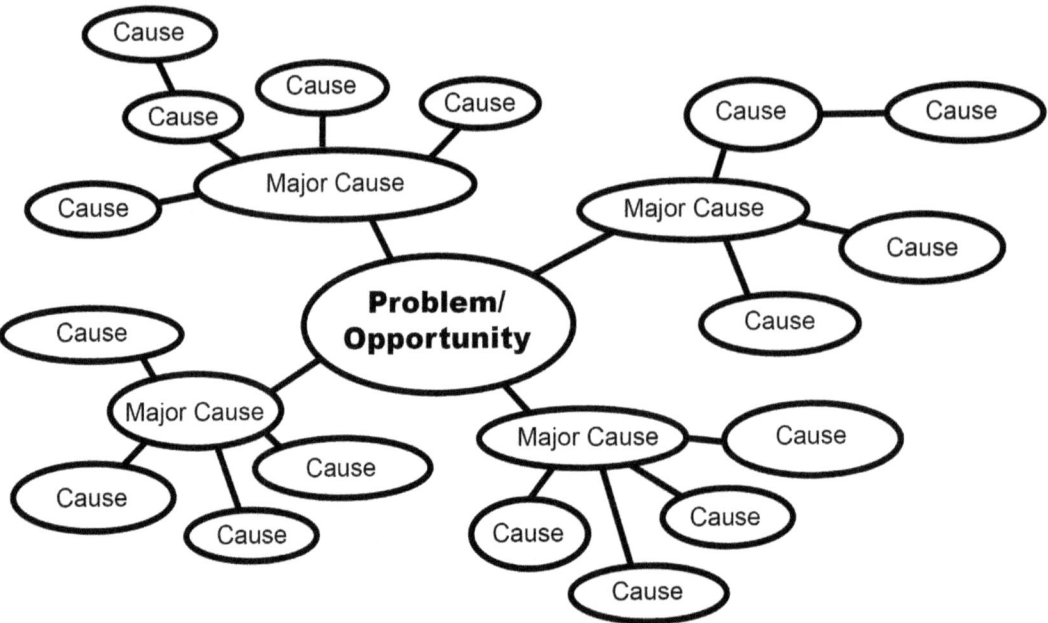

Another popular business tool proposal writers use is an Ishikawa Fishbone Diagram like the one shown in Figure 5.3. A fishbone diagram sorts causes into six categories: personnel, materials, methods, measurements, environmental factors, and machines. This type of diagram is

64

designed to help you do a root-cause analysis to determine the reasons why the problem exists. For each of the "bones" in the diagram, you can identify any causes related to the six categories.

Figure 5.3: Ishikawa Fishbone Map of Causes

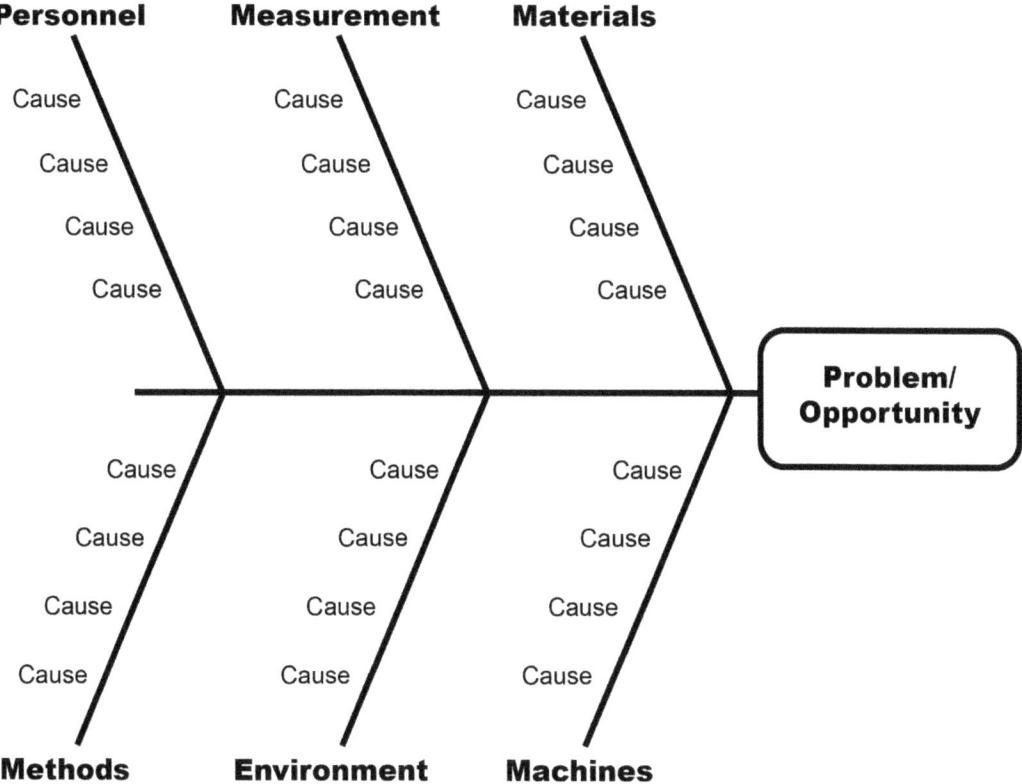

The six categories are loosely defined, but you can use the following questions to help you with your analysis:

Personnel—Are qualified and able people handling the problem? Do the managers understand what is happening and what might be going wrong? Is the labor properly trained? Do they have enough time and training to handle the work assigned to them?

Materials—Are the materials used to make the product or provide the service appropriate? Do they need to be of higher quality? Do they need to be less expensive or acquired from a better source?

Methods—Is the process for making the product or delivering the service inefficient and badly designed? Are there too many steps? Are the appropriate safety measures in place?

Environmental Factors—Are weather conditions causing problems with manufacturing or service? Are employees affected or being impeded by their work environment?

Machines—Is any equipment breaking down or obsolete? Are computer operating systems and software being kept up to date?

Measurement—Are appropriate quality control measures in place? Are methods in place for assessing whether the product or service is meeting customer needs? Do any specifications need to be more exact or perhaps less exact?

A fishbone diagram can be helpful because it urges you to see the problem/opportunity from six different perspectives. You shouldn't feel compelled to look for causes in all six categories. Instead, you can leave some of the fishbone diagram empty. You don't want to raise issues that aren't relevant to the problem or, worse, invent causes that don't exist!

Guideline 2: Ignored Problems Tend to Grow Worse

Now, it's time to map out the effects of the problem. When your readers are faced with a description of their problem and its causes, they may be tempted to minimize what is happening. They may even convince themselves that the problem will resolve itself. The interesting thing about change is that problems tend to grow over time and rarely get better if ignored. To help your readers understand the gravity of the problem, you should explain the negative effects that are currently happening. Then, show how these effects will grow worse if the problem is ignored.

Similarly, if you are trying to persuade the readers that an opportunity exists, you may use the Background section to explain what happens if the readers miss the opening. Surely others, including your readers' competitors, are also seeing the opportunity. To show what happens if someone else moves first, you can explain the effects on the readers' business or organization.

Concept mapping can be used to help you analyze the effects of a problem or a missed opportunity. Again, put the problem in the center of a page or screen (Figure 5.4). Ask yourself, "What are the 3-5 major effects of not addressing this problem?" Then go a step further and ask, "What are the 3-5 further effects that happen with each of these major effects?"

When mapping out the effects, you don't need to describe the situation in an overly catastrophic way. You don't want to give the readers the impression that the problem is beyond fixing or that the opportunity has already passed. After all, no one wants to put money into a hopeless project. You also shouldn't fabricate or exaggerate effects that are not plausible. Exaggerating the effects of a problem can cause people to stop trusting you, your company, or your organization. Instead, use reasonable and well-evidenced effects to create a sense of urgency so the readers feel compelled to take action right now.

Figure 5.4. Mapping the Effects of a Problem

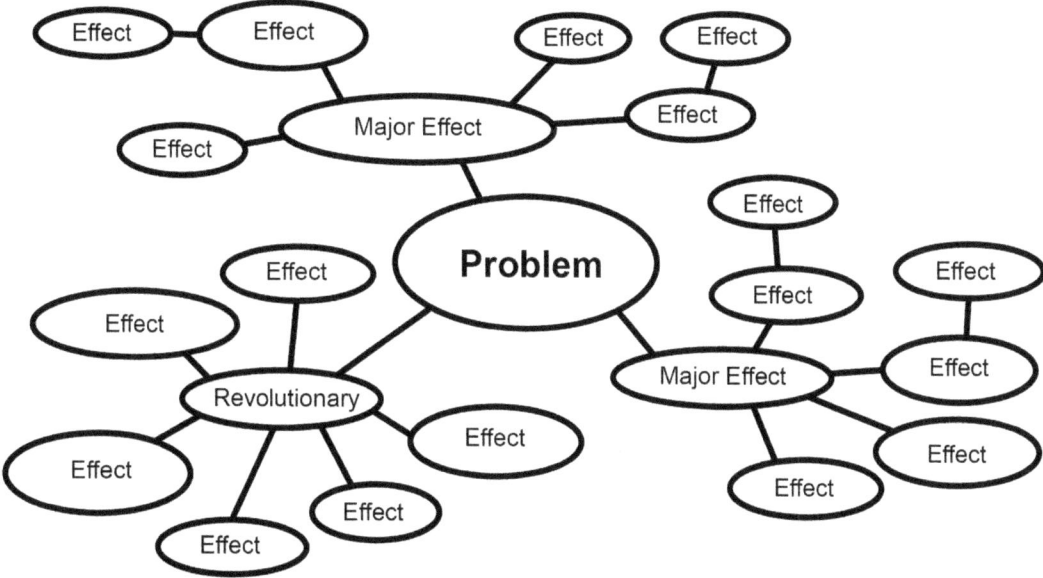

Guideline 3: Blame Change, Not People

As much as possible, you should blame change rather than people for the problem. Yes, one or more people may have screwed up, and maybe they deserve to be blamed. As you map out the causes and effects of the problem, you may realize someone made a bad call or went the wrong direction. But blaming people in a proposal, even if they are your competitors, is usually a bad idea.

Think about it. Your readers themselves may be partly at fault. Maybe they messed up, or someone they hired didn't work out. So, imagine the sour taste in your readers' mouths if a proposal stated, "If your chief engineer, Steve Wendell, had chosen to pay for regular servicing of the GH-7000 robotic arm, then your company would not be looking to purchase a new machine right now." Even if this statement is true, Steve Wendell won't be eager to accept this proposal, and neither will his supervisors. Accepting the proposal would be an admission that Steve and his supervisors had made a serious mistake.

Similarly, with grant proposals, it's usually a bad idea to blame people who need help. Yes, people sometimes make bad choices (don't we all?), but they made those decisions in complex situations. With hindsight, maybe they could have made choices. Maybe other factors that couldn't have been anticipated or foreseen may have been involved. Perhaps they knew all their options were not ideal, so they made a choice that seemed like the best available path.

In almost all cases, your Background section will be stronger if you blame change, not people. Change is ultimately the one aspect of our lives that we can do little about. Roads eventually crumble, machines break down, markets shift over time, and companies alter their strategies. Change is the culprit behind most of these things, and it will not be offended if you put the blame on it.

Mapping in Teams

Concept mapping is especially helpful when working in a team, especially on large projects. As you probably know, it's hard for groups of people to write a document together, sentence by sentence. However, multiple team members can easily participate in mapping the causes and effects of the problem.

When working with a team, find a whiteboard, chalkboard, or large sheets of paper. With a marker in hand, you can encourage the team to brainstorm about the causes and effects of the problem. As they explore the problem, write down their comments and fill the board or paper with their ideas.

The mapping process is a visual way to bring out your team's ideas. You will find yourself and your team coming up with ideas you never would have considered if you had been trying to write the document together sentence by sentence. Mapping also allows your team members to visualize the overall problem or opportunity analyzed in the Background section. It gives them an overall understanding of the changes in the current situation the proposal is trying to manage.

Researching the Current Situation

Once you have completed your concept maps of the problem's causes and effects, it's time to do some research. Your maps will point your research in the right direction by highlighting the logical relationships that structure the problem or opportunity. Mapping helps you identify the if-then, either-or, and cause-effect relationships that you can use to reason with your readers. Mapping also tends to reveal possible examples, similar cases, or anecdotes that can be used to fill out the argument in a Background section.

When researching, you should draw from a variety of sources for information and data. A good approach to research uses *triangulation* to cross-reference sources. In general. triangulation in research involves using at least three different kinds of sources related to the subject matter. These sources might include the following:

> **Electronic sources**—websites, Internet searches, electronic mailing lists (listservs), scholarly databases, government websites, television, radio, videos, and blogs
>
> **Print sources**—magazines, newspapers, academic journals, books, government publications, reference materials, microfilm/microfiche

Empirical sources—interviews, surveys, experiments, field observations, ethnographies, case studies

To write an accurate Background section, you should draw information from high-quality sources, such as government white papers, scholarly articles, or reference materials. Relying exclusively on one source—especially electronic sources that are not peer-reviewed—can be risky because you may only see a limited view of the problem you are trying to solve. Your limited choice of sources may cause a biased perspective on the topic. Moreover, your readers will certainly do their research, so you want to ensure that the sources you cite are credible to them. Most readers won't consider unevidenced output from artificial intelligence applications credible.

Solid research is the backbone of any Background section. Mapping may help you highlight logical relationships behind the problem or opportunity, but your research will support those logical arguments.

Using Artificial Intelligence to Generate Content

Artificial intelligence applications will improve over time, but they currently have limited abilities to generate the content of proposals, including the Background section. Usually, the text generated by AI applications is rather generic (e.g., "Homelessness has been a problem since the dawn of human civilization."). That generic statement will almost certainly tip off your readers that an AI was overly involved in writing the proposal, and they will assume your analysis is not specific to their needs.

We suggest using AI applications to gather and synthesize existing information available in electronic and print forms. For example, you might ask an AI application to "Identify the top five causes of homelessness in Sanders County." The AI algorithm will probably identify the causes of homelessness you already knew about, but it may also come up with one or two more that you hadn't thought of. Then, ask it to identify a few effects (e.g., "What are some of the negative effects of homelessness in Sanders County?) Again, the algorithm will probably come up with some rather generic answers, but it may also highlight one or two that you didn't expect. If the AI has an integrated Internet search function, you should ask it to provide links to relevant sources. Also, verify the quality of its information by checking its output against credible sources in your industry or research discipline.

You can then probe further into these causes and effects. For each major cause, ask what's causing it. Then, ask what are the effects.

AI applications can also generate a history of a problem you are trying to solve. You probably already know most of the major events that brought about the problem or opportunity. But there may be a few events you didn't know about or forgot about. AI applications are good at digging up information that humans may overlook. But if your proposal deals with changes in the very near past, you should check that the AI's training data covers the timeframe you're writing about. Using outdated information in a proposal can hurt your credibility with readers.

Drafting the Background Section

When writing the Background section of a proposal, you will transform your maps and research into readable sentences and paragraphs. Like most large sections of a proposal, the Background section tends to include three parts: an opening, a body, and a closing (Figure 5.5). In other words, each section of the proposal is like a miniature essay with three parts: an introduction, a body, and a conclusion.

Opening

The opening paragraph or paragraphs set a context for the section by telling the readers a few important things up front. Directly or indirectly, the opening will tell the readers the following:

- The topic of the section
- The purpose of the section
- The main point of the section

Usually, the Background section only requires a one-paragraph opening. Try to keep the opening of each section as concise as possible.

Body Paragraphs

The body of the section is where you will provide the majority of the details you mapped out in your causes map and effect map. Three approaches to writing the Background section are most effective: the causal approach, the effects approach, and the narrative approach (Figure 5.5).

The best approach for organizing the body of the Background section depends on the problem you are trying to describe and the readers to whom you are describing it. The causal approach is most effective if you need to educate your readers about the issues that caused the current problem or opportunity. The effects approach works best if the readers already know the causes of the problem and you want to explain the importance of doing something about it. The narrative approach is most effective when you want to show the readers how the problem or opportunity developed or evolved over time.

The causal approach and effects approach to organizing the Background section work best when describing the situation as it stands now. The narrative approach allows you to describe the historical events that led up to the present.

Causal Approach

As shown in Figure 5.5, the causal approach structures the body of the Background section around the causes of the problem. When using this approach, each major cause will typically receive one or more paragraphs. Essentially, the causal approach focuses on explaining the causes of the problem. Then, the closing paragraph offers a brief discussion of the effects.

Figure 5.5: Three Approaches to Writing the Background Section

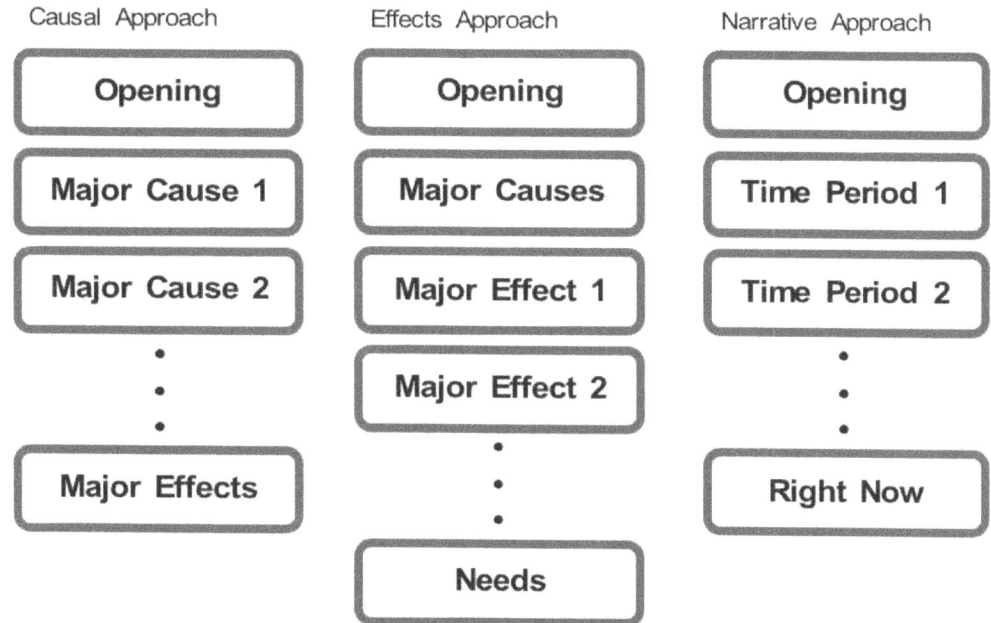

Effects Approach

The effects approach structures the section around the effects of the problem (Figure 5.5). Usually, this approach is most effective when you and the proposal's readers already know what is causing the problem. You can briefly review these causes in one paragraph, but you don't need to explain them in depth because everyone already agrees they exist. This allows you to spend more time describing what will happen if the problem is ignored.

When using the effects approach, give each major effect one or more paragraphs. Stress what will happen if no action is taken to deal with the problem.

Overall, the effects approach is most successful when your readers know why the problem exists but are reluctant to take action. By devoting most of the Background section to a discussion of the effects of the problem, you will urge them to face reality.

Narrative Approach

The narrative approach tells the readers a story about what "changed" to cause the problem or opportunity. Each paragraph moves the readers sequentially along a timeline, showing how changes within and around the company or organization require a response (Figure 5.5).

For example, let's say you are trying to convince a local business to hire your accounting firm to handle their taxes. When your potential clients started the business, their company was small, allowing them to do their accounting in-house. Ten years later, they are now a multimillion-dollar corporation, and their in-house accounting methods aren't getting the job done and causing mistakes to happen. With the narrative approach, you can start at the beginning with the founding of the company and then lead the readers toward the present, showing them how their growth over the years has led to their need for the more sophisticated accounting services that your company provides.

The narrative approach is especially useful when the readers don't fully understand why the problem exists. This approach allows you to tell a story of change that praises the readers for their success. Now, you might say, the company's success means they have outgrown the ways they did things before. It's time for a change.

Closing

The closing of a Background section can be written in a few different ways. Usually, the best closing restates the main point of the section while creating a transition into the next section (typically the Project Plan or Research Methods section). Above all, the closing should be concise, so your readers don't lose momentum before considering your project plan.

The closing in a Background section is a good place to stress the importance of the problem or opportunity. If you used the causal or narrative approach to organize the body of the Background section, you might use the closing to summarize some of the effects of not taking action. This kind of closing offers an opportunity to stress the importance of the problem, giving the readers extra incentive to pay attention to your project plan.

If you are using the effects approach to organize the body, you might discuss what the readers might "need" to solve the problem. By discussing what is needed, you will build a transition to your discussion of your plan in the Project Plan section.

Some proposal writers do not include a closing in their Background section, preferring instead to roll straight into their Project Plan section. Shorter proposals can certainly do without closing paragraphs because they tend to repeat what was just said in the body of the section. The Background section in a larger proposal usually benefits from a closing paragraph that rounds off the section and reinforces the section's main point.

Special Case: Research Grants and Literature Reviews

Research proposals, especially ones written to secure grant funding, require you to think differently about describing the current situation. In research proposals, this part of the proposal is usually referred to as the Background, Research Problem, or Literature Review. This section is designed to meet three specific goals:

- Demonstrate how the proposed research adds to or differs from prior research on the topic.
- Explain the significance of the research and its potential impact.

- Establish the credibility of the study and investigators by showing familiarity with the field's ongoing conversation on the research topic.

To meet these three goals, the Background section of a research proposal needs to summarize the previous research on the topic while identifying a *gap in knowledge* or *raising questions* about prior results. In most cases, this section of the proposal will include two types of information: (a) a review of the existing literature on the topic and (b) a summary of the principal investigator's prior research into the topic.

Literature Review

A literature review aims to familiarize the readers with the topic and the published research that has already been done on it. Literature reviews tend to be written in one of two ways: (a) a historical overview of how the field has evolved or (b) a summary of the different approaches or trends in the field:

- When describing how the field has evolved, divide the published research into 2-5 eras. Then, walk the readers era-by-era through the literature, describing how one discovery has led to other discoveries. Your goal in this kind of literature review is to show how your research project will add to the existing knowledge base, fill in a knowledge gap, or resolve a conflict in the literature.
- When summarizing the various research approaches or trends in the field, you should start by dividing your field into two, three, or four areas. After a brief opening paragraph in which the various areas or trends are named and defined, you should summarize the major published works in each. This kind of literature review aims to highlight the knowledge gaps and/or inconsistencies in the published work.

Regardless of how you organize your literature review, the best literature reviews tell a story or make an argument. If possible, you want to avoid marching your readers through a lifeless summary of the articles and books on the topic. Instead, your literature review should tell them an exciting story of discovery that has created the research opportunity you want to explore. Or, you should show them that the factions in your field have been engaged in an interesting debate that your research will try to resolve.

Keep in mind that a well-structured and well-researched literature review will go far in helping you establish credibility with your readers.

Prior Research

In addition to the literature review, you should also describe any prior empirical research you or your team have already done on the topic. Describe any analyses, experiments, or observations you have completed, showing your results. Your aim is to demonstrate that you are a competent researcher who has already gathered some intriguing data or information. You want the funding agency to see that you are on the right track; now, you need additional funding to expand your research or take it to the next level.

One thing to avoid is giving the impression that you have already figured out all the answers or solved the problem. Funding sources are most interested in supporting research that discovers something new. So, you need to give the reviewers the impression that your prior research is promising but has not yet answered all the questions. You should mention any publications from your previous research, such as refereed conference proceedings or scholarly articles.

At first glance, Background sections in research proposals might seem different from those in other kinds of proposals. But as you look closer, the similarities become more apparent. A research proposal describes how you will solve a problem like most proposals. The problem being solved, however, is a gap, unknown, or inconsistency in your field's current knowledge about the topic.

You want to show how your field has evolved, creating new problems or opportunities for groundbreaking research. Perhaps a recently published article exposed a gap in the knowledge base. Or, perhaps a new discovery threw the results of prior studies into doubt. As in other kinds of proposals, you are telling a story of change.

A good way to end the Background section in a research proposal is to stress the importance of your research. Leave any modesty aside and tell the readers how your research, if successful, will impact the field or the lives of others. If possible, quantify how many people will be affected by your research.

Chapter Summary and Looking Ahead

The Background section is sometimes a neglected part of a proposal because writers mistakenly assume the readers already understand the current situation. This is often not true.

Readers need a well-written, well-reasoned Background section for two reasons. First, they often do not fully understand the problem or the available opportunity. If they did, they wouldn't look to people like you for help. They hope you will provide them with the insights they lack by explaining the causes of the current problem and its potential effects. If they agree with your assessment of their situation, they are more likely agree to your project plan.

Second, even if your customers or clients understand their problem, its causes, and its effects, they still want to see that you fully understand the current situation. Demonstrating your airtight understanding of the current situation will go far in helping readers to trust the plan or research project you are proposing. Readers usually don't want the one-size-fits-all, cookie-cutter plans forwarded by some large consulting firms. Almost all customers and clients believe they are unique, and so are their problems. So, a well-written Background section shows the readers that you are addressing *their* problem, not fitting their problem to a predetermined solution.

In many cases, a well-written analysis of the current situation is the difference between success and failure when bidding for a project or seeking funding for research.

Try This Out!

1. Find a problem on your campus, workplace, or community that you believe needs to be addressed, such as commuting, health, or safety. Using the mapping techniques discussed in this chapter, map out the causes of that problem. Then, map out the effects of not taking action.

2. For the problem you mapped out in Exercise 1, outline three Background sections that follow the three approaches discussed in this chapter (causal, effects, narrative). Which approach would be most effective for the readers of this Background section? How does your choice of approach influence how the readers will respond to your topic and purpose?

3. Following your work in Exercises 1 and 2, write a two-page Background section in which you describe the problem, its causes, and its effects.

4. Locate a proposal on the Internet that includes a substantial Background section. How do the writers use logical reasoning or examples to support their arguments? Identify specific examples where the writers use reasoning (if-then, cause-effect, either-or). Identify different kinds of examples used in the document. Can you find places where the writers need to use better reasoning to support their points? Where might some examples help illustrate their claims?

5. Using the proposal from Exercise 4, recreate the problem-cause and problem-effect maps that may have been used to write these sections. In other words, place the problem in the middle of a sheet of paper. Then, map out the causes you see in the Background section of the proposal. Map out the effects that the proposal identifies.

6. Look at the food choices available on your campus or at your workplace. Write a one-page Background section in which you describe the current status of the food (healthy or not) where you learn and work. What is the problem with the current food choices? What are some of the causes and effects of that problem? Remember to blame change, not people, in your description of the current situation.

Case Study: The Carbon Neutral Campus Project—What is the Problem?

To this point, Anne, George, Calvin, Karen, and Tim had already defined the rhetorical situation for their grant proposal and figured out the major moves in their introduction. They had also explored the contextual factors connected to the Carbon Neutral Campus Project, including financial, social, and political pressures. They now had a much clearer understanding of their proposal's topic, angle, purpose, readers, and the contexts in which it might be used. Based on last week's meeting notes, the team began developing the proposal's Background section.

They started using a concept map to explore why the Durango University campus emits so much carbon (Figure 5.6). The team had already decided that the problem was that the campus infrastructure had been designed to only use nonrenewable forms of energy, originally coal and now natural gas and gasoline. Placing the problem "reliance on nonrenewable energy" in the middle of a sheet of paper, they began mapping out the causes of that problem.

Figure 5.6: Problem-Causes Concept Map

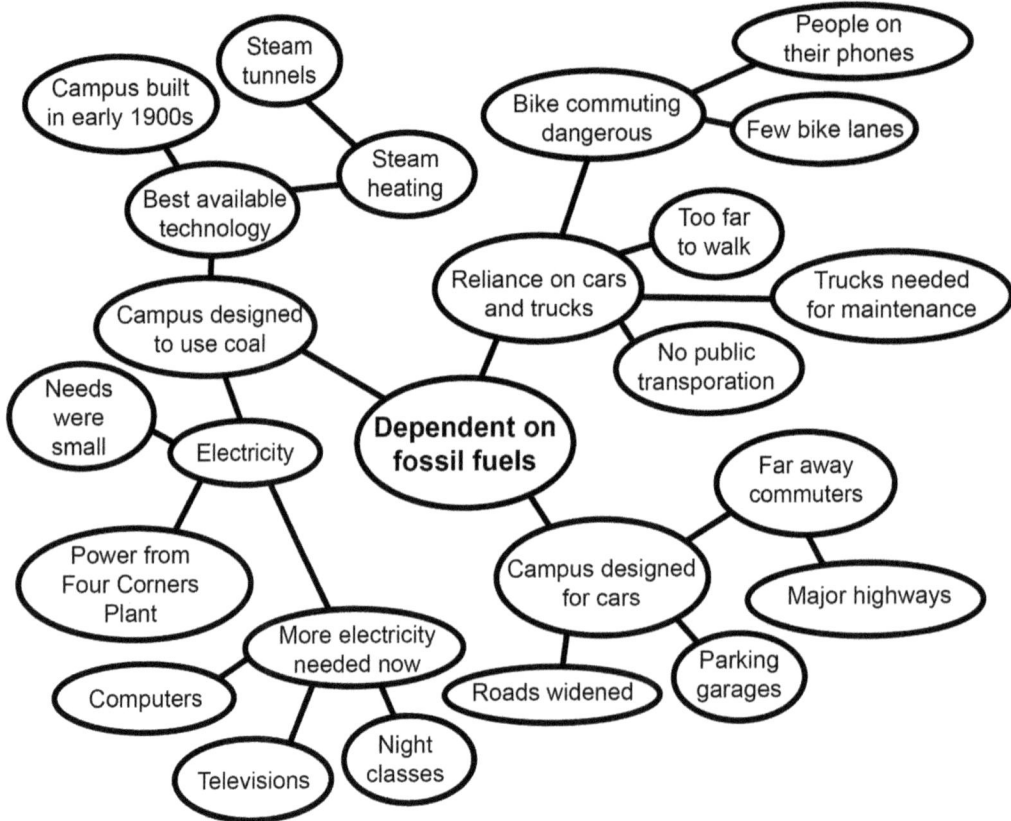

"One cause is that the campus was designed to run on coal," Calvin said. "About 20 years ago, we switched to natural gas. That was an improvement, but we're still dependent on fossil fuels."

Karen added, "Meanwhile, most people need to drive their cars to campus. This campus was designed for cars first and other kinds of transportation a distant second or third."

Tim said, "Yeah, some of us bike commute to campus, but it's dangerous out there. The major streets around campus are busy, and we have only a few bike paths. As a bike commuter, I always need to watch out for cars."

Calvin said, "The limited public transportation, too, is a problem. People who live more than a mile from campus need to go out of their way to use the bus. And the train doesn't come to campus."

"Those aren't easy things to fix," said George. "You're talking about needing to completely redesign how this campus and maybe even the city works."

As the team threw ideas out there, Anne filled out the problem-causes map on a whiteboard (Figure 5.6). She wrote down "campus designed to use coal" and then "reliance on cars and trucks" and "campus designed for cars." Then, she wrote and circled "electricity" and then "bike commuting dangerous" and "no public transportation." She began linking the circled words and phrases together.

Karen was watching the map evolve. "Let me take a guess about the underlying cause of our problem. The bottom line is that the infrastructure here at Durango University forces people to emit carbon by burning fossil fuels. They don't have a choice."

The team mapped out those problems further. When they were finished, Anne looked at the map and asked, "All right, what's missing here?"

"Money!" George declared. "Big changes will cost a lot of money. The first question many readers will ask is, 'Who's responsible for paying for all this?'"

Tim added, "Yeah, we can all agree that the costs will be the major issue with something like this. What's interesting, though, is that fixing the campus infrastructure will help the university save money in the long run by cutting energy costs and avoiding any future carbon taxes. So, a major part of our proposal might be to explain how investing money to change the campus now will save money in the long run."

"Good point, Tim." Anne said, "But let's not get too far ahead of ourselves! Let's figure out what's causing the problem before we start talking about solutions."

Tim smiled, "Sure. Sorry about that. OK, but can we agree that the overall cause of the problem is a lack of money? And, the longer we wait to do something, the more it will cost."

After finishing the problem-cause map, they began creating a problem-effects map, shown in Figure 5.7.

George asked, "OK, what are the effects of not doing anything about the problem?"

Anne said, "Well, one effect is that the Young Power Plant, which once ran on coal and now runs on natural gas, will not be around much longer. It's at the end of its life cycle right now, and no one wants to put money into a technology that will be obsolete in the next decade. We're going to need to do something, no matter what."

Karen added, "Another effect is that fossil fuels are likely to become much more expensive because of carbon pollution taxes or fees."

"Absolutely," Calvin agreed. "So, a second major effect is that the campus will lose more money over the long term. Then, those costs will be passed along to students, or the university will need to do some cost-cutting elsewhere."

Figure 5.7: Problem-Effects Concept Map

Tim put up his hand. "Another effect is that the university will just look bad if we're still running on fossil fuels a decade or two from now. So many other campuses are going carbon neutral, and we don't want Durango University to look like we aren't doing our part."

Karen added, "To compete for the best students and professors, it's important that Durango University stays technologically advanced and concerned about the future."

Anne kept scribbling the team's ideas on the whiteboard. Soon, a full problem-effects map emerged (Figure 5.7). Together, the two maps described the problem's causes and effects.

The meeting was running out of time, so they agreed that the team should split up into pairs and research these causes and effects in more depth on the Internet and at the library. Anne used her phone to take pictures of the whiteboards and offered to sketch out a rough draft of the Background section. She asked the team to send her the results of their research by the end of the week.

By Friday, the team had sent her a pile of materials to work with.

Anne decided she would use a causal approach to organize the Background section. The causal approach would allow her to educate the readers about the causes of the problem while stressing the effects of not doing something about it. With some help from an AI writing assistant, she used the concept maps and the materials the team had found to generate content for the section.

When she had a very rough draft of the Background section, she put it on the team's document-sharing site. George replied with a few suggestions for improvement. He said, "It seems rough right now, and some sections feel robotic. Also, I'm not sure if the financial issues are fully addressed in this draft."

Calvin sent a reply, "Yeah, but it feels good to have something written up. We can always clean this section up when we finish drafting the rest of the proposal. At least the basic information is all here."

"Let's move on to solving the problem and drafting the Project Plan section," Anne emailed everyone. "I think we're doing great so far!"

If you would like to see how the Background section of the Carbon Neutral Campus Project proposal turned out, please go to Appendix A.

6 WRITING THE PROJECT PLAN OR METHODS SECTION

The Project Plan or Methods Section

The Project Plan section is the heart of any successful proposal or grant. This section can also be called the *Solution, Approach, Methodology, Methods and Materials, Policy,* or *Research Design,* among other names. Your Project Plan section offers a detailed step-by-step process for solving a problem or taking advantage of an opportunity. Your plan should also identify any *deliverables* you are offering, which are the tangible results of your proposed plan.

When writing the Project Plan section, the challenge you face is creating something new. In other words, the something you are proposing does not currently exist, so you need to look into the future and use your imagination to see how that something comes into being. Writing the Project Plan section might sometimes mean adapting an existing plan or solution to a new problem. In other cases, writing this section involves inventing a whole new product, service, approach, or strategy from the ground up.

This is why writing the Project Plan section can be exciting and challenging—and sometimes frustrating. In this chapter, you will learn how to avoid some of this frustration by setting clear objectives, answering the *how* and *why* questions, and writing a well-organized plan that will persuade your readers to say yes.

Setting Objectives for the Project Plan

When drafting your project plan, you should begin by defining a primary objective and a few secondary objectives.

> **Primary Objective**—Your primary objective is the main goal of the project you are proposing. In one sentence, you should provide an overview of what your plan will achieve. When writing your primary objective, use action verbs like the ones shown in Figure 6.1. Then, state precisely what you intend to do and who will benefit from the project. For example—
>
>> Our primary objective is to design and implement a sustainable tiny homes community that provides safe, affordable, and permanent housing solutions for people experiencing long-term homelessness, thereby improving their quality of life and facilitating their reintegration into our community.

Secondary Objectives—Your proposal's secondary objectives support the primary objective, often identifying major steps or phases of your project plan. Secondary objectives can be presented as a bulleted list. For example—

To achieve our primary objective, we will accomplish the following goals:

- *Secure strategic partnerships*: Forge alliances with local businesses, nonprofits, and government agencies to gather support and resources for the tiny homes community project.
- *Engage the community*: Organize community engagement sessions to gather input, ensure local support, and promote volunteerism in the project development and implementation phases.
- *Develop a comprehensive support system*: Create a network of support services, including mental health support, job training programs, and legal assistance to aid residents in their transition towards stability and self-sufficiency.
- *Monitor and evaluate impact*: Implement a sustainable framework for the ongoing assessment of the community's impact on residents' quality of life, with the aim to continuously improve the project and replicate its success in other areas.

Figure 6.1: Action Verbs for Primary and Secondary Outcomes

Acquire	Control	Locate	Reduce
Adapt	Demonstrate	Map out	Reinforce
Analyze	Define	Manufacture	Report
Assess	Develop	Maximize	Restructure
Augment	Diagnose	Measure Modify	Score
Build	Display	Move	Strengthen
Calibrate	Distinguish	Observe	Structure
Calculate	Engineer	Optimize	Solve
Combine	Estimate	Organize	Survey
Compare	Extract	Predict	Systematize
Compute	Evaluate	Procure	Test
Construct	Formulate	Produce	Transfer
Coordinate	Gauge	Rank	Transform
Compile	Implement	Recommend	Utilize
Classify	Improve	Record	Validate
Consolidate	Launch	Recruit	Verify

Artificial intelligence (AI) applications can help develop primary and secondary objectives. You can write a rough one-paragraph overview of your project plan or methodology and ask an AI to turn that description into a primary objective. Then, revise the primary objective to fit the specific needs of the project you are proposing.

With your primary objective figured out, ask the AI to generate a list of 3-5 secondary objectives for the project. Later, when you have drafted your Project Plan section, you can revise these secondary objectives to better reflect the major steps in your project.

AI can be a helpful tool for generating initial versions of your primary and secondary objectives, but don't just accept whatever the algorithm gives you. Since AI-generated objectives lean toward being generic and nonspecific, you will need to refine them to fit your project's unique context and requirements.

Using the Objectives Provided by the Customer, Client, or Funding Source

If you are responding to a request for proposals (RFP) then your primary and secondary objectives should match or at least reflect the objectives identified by the customer, client, or funding source.

> **Read the Request for Proposals (RFP) Closely**—Look for keywords like *goal, aims, targets, ends, intentions, purpose,* and *objectives*. It's a good strategy to use the terms and phrases that they use in their RFP so your proposal's objectives match what they want.
>
> **Listen Carefully to the Point of Contact (POC)**—When communicating with the POC, listen carefully for any objectives that are being emphasized or weren't mentioned in the RFP. For example, the POC might say something like, "Above all, here are the three things we need this project to do." You will want to emphasize these objectives in your proposal.

The objectives stated in the RFP and highlighted by the POC tell you what the customer, client, or funding source wants to see in your proposal. You should reflect those objectives back to them as much as possible by using their words and ideas. Doing so will demonstrate to them that you have carefully read their RFP and that your project will give them what they are asking for.

Developing Your Own Objectives

If you are writing an unsolicited proposal or an internal planning proposal, you will probably need to develop your own primary and secondary objectives. Specifically, you want to use action verbs that signify measurable outcomes (Figure 6.1).

When crafting your primary and secondary objectives, you need to figure out what your proposal's readers *value*. What things are most important to them? What are they willing to pay for or fund? Suppose you're selling an idea to a potential client or customer. In that case, you might look through their website or marketing materials to determine what objectives and values they are trying to achieve. Pay special attention to any measurable verbs they use in their materials. Measurable verbs might include words like *increase, decrease, reduce, expand, enhance, minimize, grow,* and *accelerate*. These verbs will help you understand your client or customer's key performance indicators (KPIs). Businesses, nonprofits, or government agencies use KPIs to evaluate how well they are achieving their missions. Aligning your proposed plan with your readers' KPIs will help your readers confirm that you understand their organization, industry, and goals.

If you are writing a proposal to your supervisor or executives in your workplace, you should ask them directly, "What do you need this project to do?" Listen carefully for the action verbs they use to describe the needs of the company or the organization. Pay special attention to moments when they express their personal values or needs. You can also use your workplace's KPIs to refine your proposal's objectives.

Answering the *How* Questions

With your objectives defined, you now need to begin answering the how question: *How* will you solve this problem or take advantage of this opportunity? Or, put another way, *how* will you achieve your primary and secondary objectives?

A concept map can be a helpful tool for answering the how question because it helps you visualize how your project will work and its logical structure. To begin, it is important to recognize that your project's objectives will be met when you and others take specific steps to reach them. So, when mapping out the solution to a problem, you want to identify all the larger and smaller steps that will allow your project to achieve its goals.

Step 1: Write Down A Few Solutions to the Problem

To begin the mapping process, you first need to identify a solution that might achieve your primary and secondary objectives. While drafting the Background section, you more than likely thought of a few possible solutions that might work.

On the first try, you do not need to identify the best solution. Create a list of possible solutions, perhaps using AI to generate some additional ideas for solutions. Then, you can map them out separately, seeing which one best meets your primary objective and secondary objectives.

Step 2: Map the Major Steps of a Possible Solution

As discussed in the previous chapter, mapping allows you and a team of others to figure out the logic behind your ideas. When sketching out a project plan, you can use mapping as a logical method to identify some of the major steps:

1. Place your most promising solution in the center of a whiteboard, computer screen, or sheet of paper. Circle it. (Concept mapping software applications are available on the Internet to create concept maps on your computer screen.)

2. Ask yourself and your team, "What are the two to five major steps needed to make this solution happen?"

3. Write down those major steps around the solution, circle them, and connect them to the solution (Figure 6.2).

4. Add an *analysis* or *assessment* step that will describe how the project will be evaluated or assessed. This step should explain how you will use metrics, data analysis, product testing, customer satisfaction surveys, or impartial evaluators to determine if the project met its objectives.

As a rule of thumb, limit your project plan to about 4-7 major steps. A plan that includes more than seven major steps starts to feel unmanageable to the readers. The long sequence of steps will make the project seem too complex to your readers and, therefore, less doable.

If you need more than seven steps, you can divide your project into *phases* or *stages*, each with 4-7 major steps. This will keep the project plan from feeling too complex for the readers.

Figure 6.2: Using a Concept Map to Develop a Project Plan

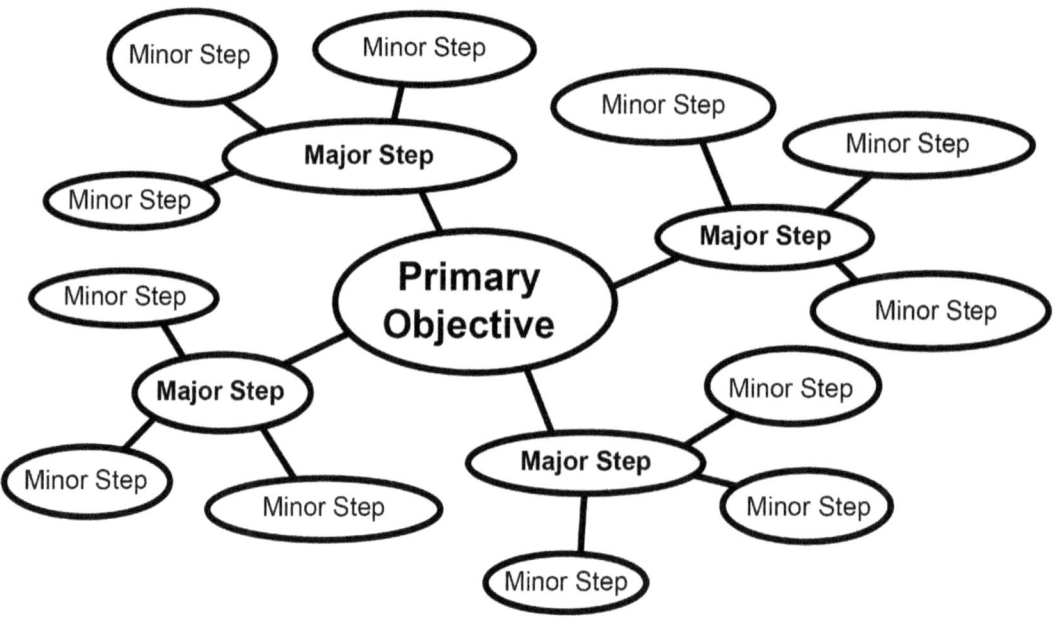

Step 3: Map the Minor Steps for Each Major Step

Then, map out the minor steps in your project plan. For each major step, ask yourself or your team to identify the 2-5 minor steps required to achieve that major step. Write those minor steps around each major step. Circle them and connect them to their major step (Figure 6.2).

When mapping out the minor steps, keep asking the *how* question. How can we achieve this major step? How will we accomplish each of these minor steps? As you map further out, you will see your project plan developing into something more detailed and concrete. As you identify the minor steps and tasks within the project, your plan's broader major steps will be broken down into specific actions and tasks to be completed.

An Ishikawa fishbone diagram can be a helpful tool for seeing your project plan from multiple perspectives. As shown in Figure 6.3, Ishikawa's six categories can be used to identify the steps needed to achieve the project's primary objective.

The bones in the fishbone diagram allow you to analyze the project plan's steps from six different points of view. A fishbone diagram can help you discover steps you might have overlooked if you didn't view your plan from other perspectives. Here are some questions you can use when filling out the fishbone diagram for your project plan.

> **Personnel**—Who are the best managers to handle the project? Can they be pulled over from other projects if necessary? What kind of labor do you have available? Who will you need to hire? What hiring and training will be needed to prepare your workforce to take on this new project?
>
> **Materials**—Where will you need to source materials or supplies for the project? What suppliers or contractors can provide you with the necessary quality at a reasonable price?
>
> **Methods**—How might you retool your production line or rethink how your service is delivered? What safety measures and quality control procedures will need to be put into place? Will consultants be required to train employees in these new methods?
>
> **Environmental Factors**—How might ecological factors and weather conditions impact the project? Are there any social or political concerns that might influence or undermine what you are trying to do? How will environmental laws impact the project?
>
> **Machines**—What equipment or computers will you need to purchase or refurbish to complete this project? Would more advanced machines or computers work better, or do you already have the appropriate equipment to handle the project? Do you have the parts, lubricants, cleaning supplies, and personnel to keep your machines and computers up and running?
>
> **Measurement**—How can you measure the quality of the finished project? How can you assess whether you are meeting the customers' expectations for the finished product or service? Have you described appropriate places and times to test products in your project plan?

Exploring these six categories will help you generate additional questions that your project plan should answer. As you fill out the fishbone diagram, keep in mind your customer's or client's unique needs, values, and tendencies.

Figure 6.3: Using an Ishikawa Fishbone Diagram to Identify Steps in a Project Plan

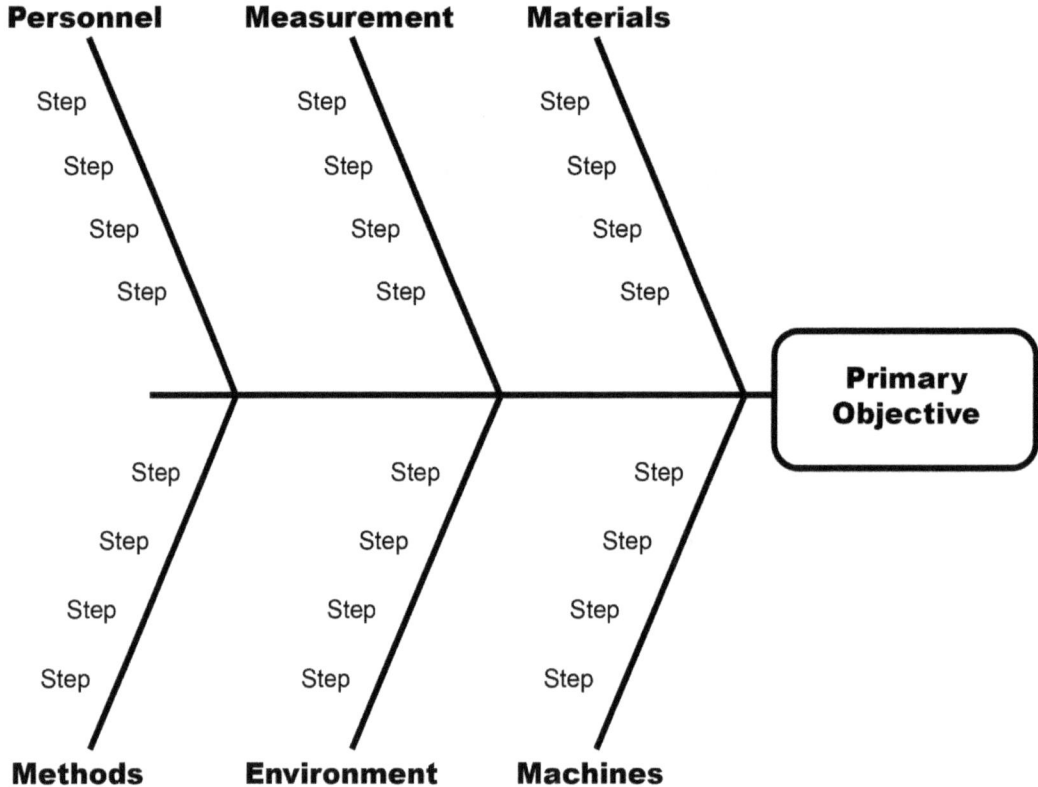

Step 4: Determine if Your Plan Can Achieve the Project's Objectives

When you are finished mapping, review the primary and secondary objectives you identified earlier. Ask yourself whether the project plan you have mapped out can meet these objectives. Are there any objectives, especially secondary objectives, not met in your concept map or fishbone diagram?

When reviewing your objectives, you may find yourself adding additional major and minor steps to your map or fishbone diagram. In some cases, you may cross out major or minor steps because they are outside the scope of your objectives.

You may even find that the steps in your map don't achieve your primary objective or secondary objectives. If this happens, try mapping out one or two other solutions to see if another approach would better achieve your objectives.

In the end, mapping will allow you to answer the how questions about your project. Your map of the project plan illustrates roughly how you will achieve the primary and secondary objectives you identified, thus solving the problem or taking advantage of the opportunity.

Step 5: Answer the Readers' *Why* Questions

When reviewing your proposal, your readers' overriding question will be, "Why?" Why should we do it this way? Why are these major steps necessary? Why couldn't the project be completed a different way? As the proposal writer, your job is to answer these why questions as you explain your plan's major and minor steps.

Think of your answers to the how questions as the skeleton of your plan and the why questions as the ligaments that hold your plan together.

A good way to develop answers to your readers' why questions is to use a How, Why, What table like the one in Figure 6.4. Write down one of your major steps on the top line of the table and state a brief answer to the why question (i.e., "Why is this major step needed?). Then, list the minor steps that will happen as you complete this major step and answer the why question for each of them. For each minor step, ask yourself, "Why is this step needed?"

When you write your project plan, your answers to the how and why questions will work hand in hand. For each major and minor step, you will essentially tell the readers, "This is how we will do this step, and this is why we will do it that way." This back-and-forth between how and why will create a powerful step-by-step description of your solution to the problem.

Step 6: State Deliverables for Each Major Step

When you have answered the why questions, ask yourself what deliverables will be produced in each major step. Deliverables are the measurable results or outcomes for each major part of your project plan. Think of deliverables as the things you will "deliver" to your readers as the project progresses and when it is completed.

In some cases, deliverables are finished products (e.g., a machine, a building, software, a training module, a renovation). In other cases, a deliverable might be some form of communication to the readers (e.g., a progress report, a completion report, a research article, or a presentation).

Whenever possible, you should try to come up with some kind of deliverable for each major step. After all, your customers and clients will want to observe your work's tangible results as you progress. That way, they will feel they are receiving something they can see or touch in return for their investment in your project.

In grant proposals, identifying deliverables is one of the tricks of the professional grant writer's trade. Grant writers know that funding sources don't like to fund projects that go silent after the funds are sent. They are also cautious of funding projects that look like they will shut down when the funding runs out. They like to see that the project will continue into the future, even after they step back from funding it.

By demonstrating that each major step in the project leads to a deliverable—even a regular progress report—you can reassure your readers that the project will keep progressing and that their funds are leading to tangible results. You are also showing them that the project will continue into the future, even after their funding runs out.

When you finish filling out the How, Why, What tables for each major step, you will have sketched out the content of your Project Plan section. You can now turn each of these tables into one or more paragraphs in your Project Plan section. Essentially, the paragraphs built from these tables tell the readers *how* you are going to achieve a particular part of the plan, *why* you think these major and minor steps are needed, and *what* deliverables will result as the project progresses.

Of course, you will still need to track down sources, statistics, or facts that support your answers to the why questions. Your research for the Introduction and Background sections should already provide some leads, and an AI with integrated Internet search functions can further help you find published support for your approach. You are ready to start drafting your Project Plan section when you have finished filling out your How, Why, What tables.

Figure 6.4: Using a How, Why, and What Table for Major Steps

Major Step:	
Step 1:	Why?
Step 2:	Why?
Step 3:	Why?
Step 4:	Why?
Deliverables:	

Drafting the Project Plan Section

In Chapter 5, you learned that each section in a proposal typically has three parts: an opening, a body, and a closing. All three parts should be included in a well-written Project Plan section. An opening is needed to tell the readers your project's purpose, objectives, and main point. The body of the Project Plan section will describe the steps in your plan. The closing will round off the Project Plan section by summarizing the deliverables you are promising readers. Figure 6.5 shows how this section tends to be organized.

The Opening of the Project Plan Section

The opening of the Project Plan section is designed to forecast the information in the body of the section. The opening of the Project Plan section is especially important because it is where your readers need to make the critical transition from your description of the current situation to your description of the plan.

This transition can be difficult because the Background section might have set a negative tone. After all, in your Background section, you just told the readers that they have a problem and that some negative consequences will occur if they don't take action. Even in a Background section that describes a golden opportunity, you will probably close your discussion by mentioning some of the negative effects of not taking advantage of the opportunity.

The Project Plan section, on the other hand, needs to be as positive as possible. From this point forward in the proposal, you will concentrate on the advantages and benefits of solving the problem.

Essentially, your opening paragraph in the Project Plan section should pivot the readers from a negative viewpoint to an optimistic one. To negotiate this tricky transition, the opening of the Project Plan section should make most, if not all, of the following important moves:

Transition—signals to the readers that you are starting your discussion of the plan

Identify and Name Your Overall Solution—in one sentence or phrase identify and name your overall strategy for solving the problem

Statement of the Purpose of the Section—tells the readers that the purpose of this section is to provide a detailed step-by-step plan

Statement of the Plan's Objectives—lists the primary and secondary objectives that any successful plan would be able to meet

Forecast the Plan's Major Steps—briefly forecast the major steps of your plan

Of course, these moves do need not to be made in this order, and you don't need to make all of them. However, if the opening of your Project Plan section can account for these moves, you will have established a clear framework for the detailed plan that follows.

Figure 6.5: The Major Parts of a Project Plan Section

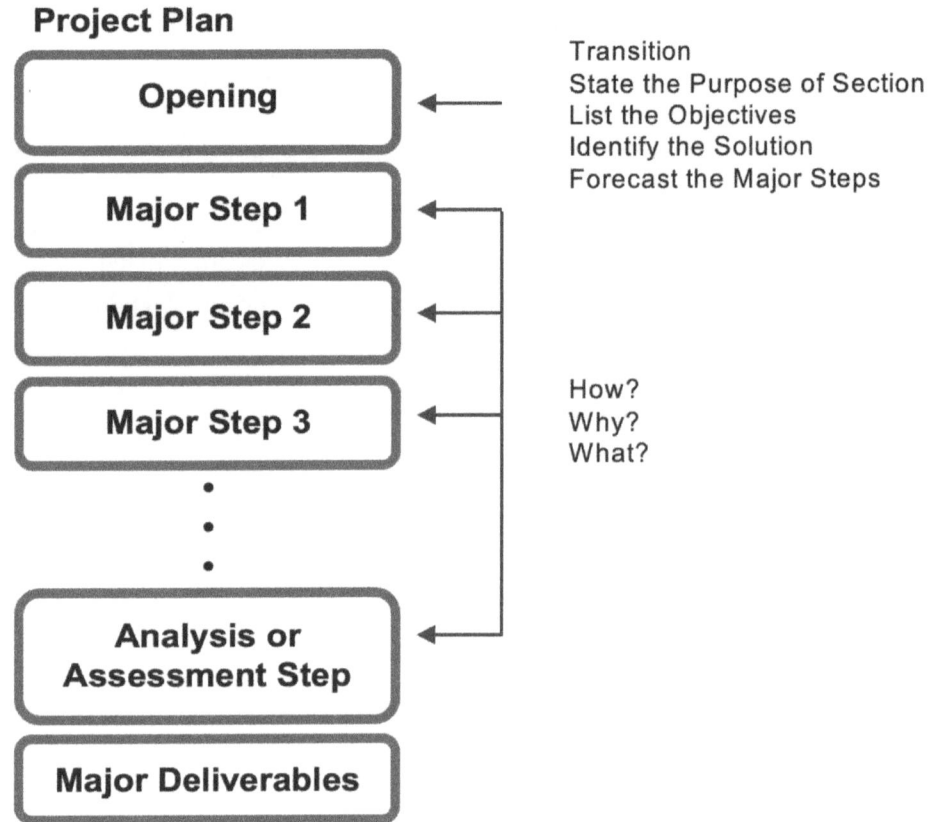

The Body of the Project Plan Section

The body of the Project Plan section is where you are going to explain your plan step by step and tell your readers why you believe the problem should be solved in this particular way. Your proposal's Project Plan section should be organized around the major steps you identified earlier. For example, if you have four steps in your plan, the body of your Project Plan section will likely have four major parts. Each part will describe one of the steps in detail.

Drafting each of these parts should not be difficult at this point. Start by looking over the How, Why, What table that explains the first major step in your plan. State that major step up front. Then, state why the major step is needed. And then, discuss the minor steps needed to achieve it. As you describe these minor steps for the readers, flesh out the discussion by answering the readers' *why* questions for each of the smaller steps.

After you discuss all the minor steps within each major step, you should identify the deliverables that will be produced. Deliverables, as mentioned earlier, are finished products, services, or communications you can show or deliver to the readers. At a bare minimum, you should promise a deliverable like a progress report or a completion report that shows the readers what has been accomplished at each stage of the project.

As shown in Figure 6.5, a major step you should also include is an analysis or assessment step. The purpose of this major step is to determine whether the project met the original objectives identified in the Project Plan section's opening paragraph. Specifically, you should explain how the final product or service will be assessed through quantitative or qualitative means. Where possible, align your assessment strategy with the client, customer, or funder's metrics or KPIs.

If you are writing a grant proposal, you might also tell the readers that outside evaluators or inspectors will be brought in to determine whether the project was completed properly and to the desired quality level.

The deliverable for the Analysis or Assessment step is typically a completion report produced by your company, organization, or outside evaluators.

The Closing of the Project Plan Section

In the closing paragraph of the Project Plan section, you should summarize the major deliverables you promised your customers or clients in the body of the plan. Mainly, you want to highlight the tangible results that will come about if your readers say yes to the proposal.

Keep this closing brief. It should round off your description of the plan and prepare the readers to make a transition into a discussion of your company's or organization's qualifications.

Developing a Project Timeline

Your proposal's Project Plan section should offer a project timeline. To create a timeline, each major step should identify the date it is scheduled to be completed. A timeline provides your clients or a funding source with milestones to track your progress.

To develop a timeline, start with the project's deadline and work backward to its starting date. Assign a specific date to each major step in the project plan. Specialists in time management call this "backward planning," and this process will help you scale the project's timeline to fit the completion date.

By working backward from the deadline, you will avoid the temptation to be unrealistic about how much time the later steps (and often the most important steps) will require in your project plan. Remember the project manager's idiom: as soon as a project starts, it is already behind schedule and over budget. Build enough cushion into your timeline that you can avoid getting too far behind schedule.

Even if you aren't exactly sure about dates, you should include them. Real dates will make your project seem more realistic, giving the renders a clear sense of when you expect the project to start and be completed. If, for some reason, you cannot start the project on time, the dates on the timeline can be adjusted accordingly. In these situations, you will need to negotiate the new dates with your customers, clients, or the funding source.

Larger proposals and grants often use graphics to illustrate the project timeline. You might include a list of dates and events or a chart that shows important dates and the major steps that will be completed by those dates. Gantt charts, like the one shown in Chapter 12, are especially popular in proposals because project management software programs and AI applications can create them quickly for the project.

A Comment on Research Methods Sections

If you are writing a research proposal, especially one aimed at securing grant funding, the process described in this chapter might not appear suitable for writing a Methods section (sometimes called *Methodology, Methods & Materials,* or *Research Design*). However, when you recognize that the Methods section describes a project plan, you will see that the process described in this chapter works well.

A Methods section describes how and why a subject will be studied in a particular way. This section needs to do more than describe how you will complete your study. It also needs to explain *why* each major and minor step is necessary. After all, the reviewers of your proposal will scrutinize your methods closely to determine whether your research project will yield useful results.

If you are using a methodology adapted from another study, you should carefully describe the methodology and cite it. If you have decided to invent a new methodology, you will need to justify your decision to study your subject in a new way.

In a research proposal, the Methods section often begins with a list of objectives that the study will strive to achieve. In most research proposals, this list of objectives is specific about the kinds of quantitative or qualitative results that would signal whether the research project was a success. Much like a business proposal, your research objectives establish the study's goals and identify the benchmarks for success.

If your research subjects are people, animals, insects, or places, the Methods section should be clear about their characteristics and the environments and circumstances (e.g., time, temperature) in which these subjects will be studied.

After defining the subject, a Methods section usually offers a step-by-step description of the process that will be used to do the research. Your description of the research methods should be exact and complete, mentioning any materials, tools, formulas, and calculations that will be used during the study. Essentially, the description should be precise enough that other researchers could replicate the study and calculations to test its results.

Methods sections often close with a discussion of the computer hardware and software that will be used to crunch the numbers and analyze the data generated by an experiment or observations. You should mention any statistical procedures or software packages you will use to process the data. Some researchers even mention the brands of computers that will be used to analyze the results of the study. That way, other researchers can exactly replicate the analysis of the results.

Always remember that the Methods section is one of the most scrutinized sections in any research proposal. The methods help the reviewers determine whether the results of the study will be valid and useful. If the proposal's reviewers have doubts about a study's methodology, they probably won't recommend funding the project. After all, a flawed methodology will lead to flawed results.

Chapter Summary and Looking Ahead

The Project Plan section is the heart of your proposal. It explains how the problem would be solved and why it should be solved in a particular way. Most proposal writers see the Project Plan section as the most challenging part of the proposal-writing process. Once the Project Plan section is drafted, the remaining sections tend to be easier to write.

Try This Out!

1. Find a proposal or grant on the Internet and analyze its Project Plan or Methods section. Do the proposal writers provide a step-by-step process for achieving stated (or unstated) objectives? Do they answer the *why* and *how* questions in each part of their plan? Do they identify deliverables at the end of each step or the end of the plan? Write a critique of the proposal's Project Plan section or Methods section to your instructor in which you explain why this section was effective or not. Offer recommendations for improvement.

2. Using the proposal you found for Exercise 1, reconstruct the outline the writers may have used to draft the section. Put the solution at the top of your screen or a sheet of paper. Then, outline the major and minor steps that make up their plan. Looking over this outline, does their solution seem logical and reasonable? Are there any gaps in content or organization you would like to see filled? Would you have organized the section differently?

3. Look closely at an RFP in your area of interest. With a team, identify the primary and secondary objectives that the customer, client, or funding source would like the project to achieve. Can you think of any unstated secondary objectives not mentioned in the RFP that might be important to the readers?

4. Working with a team, find a problem in your campus, workplace, or community that needs a solution. Identify some objectives that any successful solution would need to achieve. Then, map out a plan to solve the problem.

5. Imagine you have been asked to develop a mentoring program at your college or workplace. The current problem is that new students or employees often feel overwhelmed by the immediate onslaught of the work. As a result, they sometimes drop out or quit within

a couple of months. Your task is to set objectives, map out a solution, and write a two-page description of your mentoring plan. Your plan should answer the *how* and *why* questions while providing some tangible deliverables.

Case Study: The Carbon Neutral Campus Project— What's the Project Plan?

Drafting the Background section was challenging for the Carbon Neutral Campus team. However, with a better understanding of the problem, they were eager to figure out how to solve it. They felt re-energized as they shifted their focus to drafting the Project Plan section.

"OK. Let's figure out how we can make our campus carbon neutral," Tim said as the group came in the meeting room.

George picked up a marker and stood at the whiteboard. "Let's begin with the objectives. What do we want our plan to achieve? What goals should any successful plan meet?"

Anne looked at her notes, "Earlier we said that our primary objective is to develop a comprehensive strategic plan for converting the Durango University campus to sustainable forms of energy."

George jotted this down at the top of the whiteboard (Figure 6.6).

Karen added, "It's also crucial that we start conversations about the importance of energy issues on campus and in the community, especially related to the climate crisis. It's such a big abstract issue that most people can't see what we can do locally."

"I agree," George said as he wrote, "Start conversations about energy and climate crisis" on the whiteboard.

After a pause, Calvin added, "I would like to second Karen's point by stressing the importance of involving people in the local community. Speaking as a townie, I can honestly say that working with this group is the first time I've ever had discussions about environmental issues with someone from the university."

"Hmm, yes," said George as he continued writing on the board. "We need to start a dialogue between the university and the community."

Anne added, "One of my major concerns is that our plan won't be sustainable and might be set aside after a few years. You know, priorities and budgets change. We need to think about the future and create a flexible plan to guide long-term decisions."

"Good point," George agreed as he wrote. "We need a realistic and useful plan in the long run. Otherwise, all our efforts will be for nothing."

"I understand that we're thinking about our campus," Tim said, "but it would be cool if we could come up with something that would help other colleges transition their campuses to clean energy sources."

"True," Anne said. "Also, if we can determine how our approach can be applied beyond our campus, the Tempest Foundation will find it more attractive."

Figure 6.6: The Concept Map for the Project Plan Section

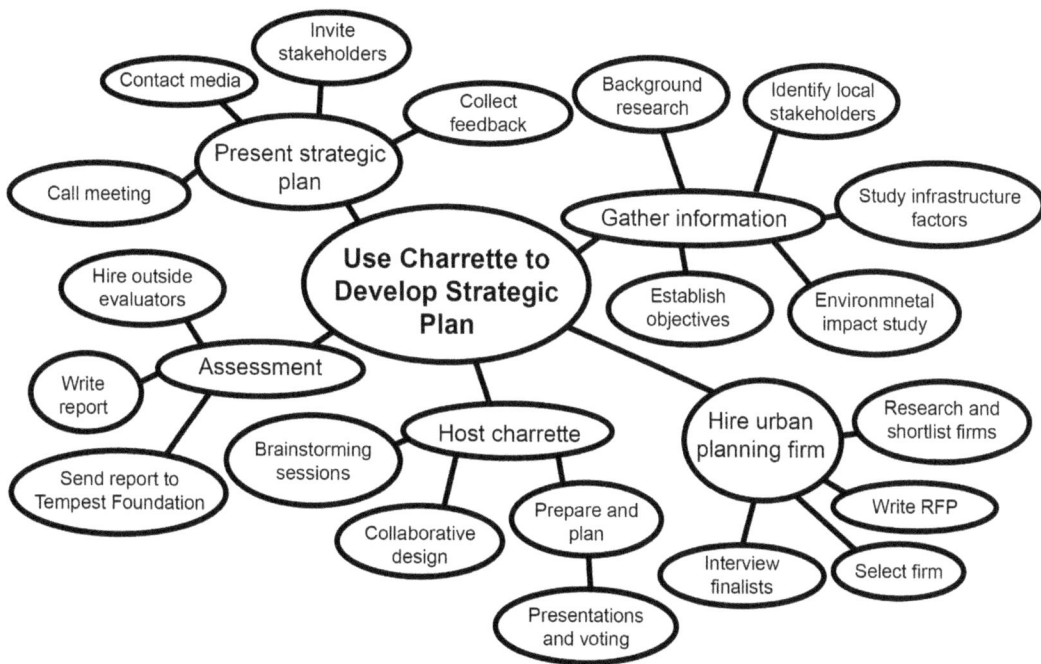

"OK, this list of objectives is looking promising so far," said George as he finished adding, "Create a plan that could be used as a model for campuses" to his list. "Now comes the challenging part. What kind of project would allow us to achieve all these objectives?"

Calvin said, "You know, I've been giving this a great deal of thought. Last year, I participated in something called a 'charrette' that used collective urban planning to revitalize a neighborhood in Albuquerque. We could probably do something like that here."

Karen looked confused. "A charrette? I've never heard of that before."

"Yeah. I hadn't heard of a charrette until I went to one," said Calvin. "It's an urban planning tool that brings all the local stakeholders into the planning process. Everyone is invited, including community leaders, citizens, administrators, customers, business owners, and politicians. They all come together for a Friday or a weekend. Then, people work in design teams to develop various plans for a neighborhood. The teams are guided by urban planners who are hired as consultants. In Albuquerque, the people who showed up were really engaged. On the Sunday of the charrette, we presented our design plans and then discussed and debated the best ideas. Then, we voted on the top plans. It's an exciting way to involve the community while getting a wide range of new ideas."

Tim asked, "OK. But what happens next? Who follows up? Who finalizes the plan?"

"That's the interesting part," said Calvin. "The urban planning firm took all those plans and comments back to their office. Two months later, they called a meeting to present a first draft of a strategic plan for rebuilding the neighborhood. It was really cool how they transformed the ideas of amateurs into professional drawings and schematics."

Anne liked the charrette idea, "I think a charrette could work. It encourages grassroots involvement. In my experience, big top-down decisions made by administrators or consultants usually get a lot of pushback from faculty, students, staff, and the local community. Stakeholders like to be involved in the decision-making process."

Karen had been skeptical, but she admitted, "I kind of like this charrette idea. It sounds promising."

George wrote in the center of the board, "Use a charrette to develop strategic plan." He circled it and asked, "All right, what are the two to five major steps we need to take to achieve this objective?"

The group started brainstorming and filling out the concept map on the whiteboard. The significant steps included "Gather information for charrette," "Host charrette," "Hire an urban planning firm as consultants," and "Present the strategic plan." The team also decided a steering committee was needed to oversee the process, including hiring the urban planning firm.

They were surprised at how quickly their plan expanded in depth and detail. Their concept map captured their responses to the how questions, showing how the plan's major and minor steps would work.

They then shifted their attention to addressing the why questions, using How, Why, What tables to fill out each major step of their project plan. George drew these tables on the whiteboard, and they filled them in, identifying deliverables for each significant step (see Figure 6.7).

The hour passed quickly, and a few team members needed to go. So, George and Calvin agreed to draft a rough version of the Project Plan section for the grant proposal. After a couple of weeks of collaboration, they emailed the team their rough draft.

"Wow," Tim responded, "This is incredible. We've made this from scratch, and it looks fantastic!"

Anne nodded and said, "That's the great thing about proposals—they allow us to be creative and come up with innovative solutions. The hard part is behind us."

Figure 6.7: A How, Why, What Table Created by the Carbon Neutral Campus Planning Team

> **Major Step:** Hire Urban Planning Firm as Consultants to Lead Charrette
>
> | | |
> |---|---|
> | **Step 1:** Research and shortlist firms | **Why?** Invite the best firms to apply |
> | **Step 2:** Send out request for proposals | **Why?** Create competitive bidding |
> | **Step 3:** Interview finalists | **Why?** Select firm that fits our culture |
> | **Step 4:** Select a finalist | **Why?** Provide winning bidder time to prepare |
>
> *Deliverables:* A recommendation report to the university administration

If you would like to see how the Project Plan section of the Carbon Neutral Campus proposal turned out, please go to Appendix A.

7 WRITING THE QUALIFICATIONS SECTION

People Hire People, Not Proposals

The Qualifications section occupies a sensitive place in the structure of a proposal. In the Project Plan section, you energized the readers by showing them how their problem can be solved or how they can take advantage of an opportunity. If you are writing a grant proposal, you just finished describing your interesting new project or explaining your research methodology. If they are still reading your proposal, they are considering your project, product, or service. Otherwise, they would have already set aside your proposal or grant.

In your Qualifications section, you need to maintain that momentum. A well-written Qualifications section will continue building the readers' interest by describing the strengths and capabilities of your team, company, or organization.

The purpose of the Qualifications section is to demonstrate to the readers that your team, company, or organization has the personnel, experience, expertise, and facilities to carry out the project described in the Project Plan section. A well-written Qualifications section should also persuade the readers that your team, company, or organization is *uniquely* qualified or the *most* qualified to take on this project. You need to show them the special qualities that set you and your team above the competition.

Keep in mind that proposals and grants build relationships. When readers accept your proposal, they are placing their trust in you, your team, and your organization. This trust is crucial. Without it, even the best plan won't get readers' approval. The Qualifications section of your proposal plays a vital role in fostering this trust. Here, you establish your *ethos*—your credibility—convincing your readers of your abilities and integrity.

Readers of proposals and grants pay special attention to a proposal's Qualifications section, so you should put your full effort into this section. Inserting a generic backgrounder about your company or organization will not get the job done! In this chapter, you will learn how to make your proposal's Qualifications section a dynamic component that builds your readers' trust and keeps them excited about the project, product, or service you are pitching.

Types of Qualifications Sections

The Qualifications section is where character (your *ethos*) takes the lead. In a proposal, character is built on the motives, values, and attitudes that your company or organization shares with the readers. When crafting this section for your proposal, start by considering the existing relationships between your company or organization and your readers' company or organization.

If your readers are not familiar with your company or organization, you should provide a detailed description of your management, labor, facilities, equipment, and relevant experience. This description must be consistent with the steps outlined in the Project Plan section. If you have a long-standing relationship with the readers' company or organization, you can usually pare the description back to a core description of your company and highlight anything that has changed or is new.

Internal proposals, a bit differently, will usually emphasize the qualifications of team members who are proposing the project because managers in the company are already acquainted with the available resources and infrastructure.

Qualifications sections can be sorted into three types: *business-to-business*, *team-to-management*, and *recommendation*. All three types are designed to show the readers *who* will be involved in the project, *why* these people are uniquely qualified, and *what* resources and facilities will be used or needed.

> **Business-to-Business**—a description of your company or nonprofit organization that demonstrates it is uniquely qualified to handle this project. You will tell the readers who will be working on the project, explain their credentials and experiences, and show what kinds of resources will be devoted to work.
>
> **Team-to-Management**—written for internal proposals pitching new ideas within the company or organization. This kind of Qualifications section is designed to show management that your team can carry out the proposal's project plan.
>
> **Recommendation**—a recommendation to hire an outside contractor, supplier, or consulting firm that can carry out the plan. Recommendation qualifications sections are often used to refer a customer or client to the best company for implementing the plan.

Keep in mind that not all proposals need a stand-alone Qualifications section. In some cases, especially with short proposals, you can describe your team's qualifications in your Project Plan section. As you answer the what, why, and how questions in the Project Plan section, you can identify who will work on the project and what facilities and equipment are needed.

What Makes You Different Makes You Attractive

In a Qualifications section, you need to prove to the readers that your team, company, or organization is uniquely qualified or best qualified to do the project. You want to show them something unique or different about your company or organization that makes you especially qualified to do the work.

Of course, you should always remember that all companies and organizations have their strengths and limitations—there is no perfect team for any project. But by paying attention to what makes your company or organization different from your competitors, you can often persuade your readers that your team has the know-how, common sense, commitment to excellence, and attitude to best meet their needs.

A good strategy is to present your differences as strengths. For example, proposal writers at small companies will sometimes complain, "Our competitors are the big sharks in this field. It's hard to compete with their huge workforce and facilities." Ironically, proposal writers at large companies, feeling more like whales than sharks, will complain that they are competing against all those aggressive smaller companies that are more flexible and can avoid the massive overhead costs of a large corporation.

The lesson is this: *What makes you different makes you attractive.* If you work for a smaller company in the industry, you should look for ways to point out that your company's smaller size allows it to be flexible and innovative—unlike your larger competitors, who will waste gobs of money on overhead and try to sell the "same old pre-packaged solution." If you work for a larger company or organization, you should stress your experience, equipment, facilities, and the ability to put the right people on the project. A strong Qualifications section capitalizes on your company's or organization's unique qualities, showing the readers why you are better than your competitors.

A Strengths and Limitations Worksheet like the one shown in Figure 7.1 can help you figure out what makes your company or organization different. Here's how to use the worksheet:

1. List your top competitors at the top of the worksheet, assuming they are pursuing this opportunity, too.

2. Write down each of your competitors' strengths and limitations. Consider issues such as size, experience, personnel, facilities, and any previous history with the customer, client, or funding source.

3. Write down your company's or organization's strengths and limitations. Be honest with yourself—you are trying to develop a candid assessment of where your company or organization fits among its competitors.

4. Circle any strengths of your company or organization that your competitors lack.

5. Transform any of your competitors' limitations into your strengths and write them down in your Strengths column.

For example, perhaps your company has a research and development (R&D) department, but your competitors don't. You might point out in your Qualifications section that your company's ability to commit full-time researchers to the project is an advantage. Also, mentioning this fact in your Qualifications section will draw attention to your competitors' absence of R&D departments.

In grant writing, identifying your strengths (and the limitations of competitors) can also be helpful. Whether we want to admit it or not, nonprofits compete against each other for funding.

By highlighting your organization's strengths, you can show that your proposed research or project is more likely to succeed than the ones proposed by your competitors.

Figure 7.1: Strengths and Limitations Worksheet

	My Company or Organization	Competitor 1	Competitor 2	Competitor 3
Strengths				
Limitations				

Artificial intelligence (AI) can be useful when researching your company's or organization's competitors. An AI application can dispassionately aggregate information on your competitors to give you a broader view of their strengths and weaknesses. AI may also offer an outsider's view of your company or organization so that you can assess your own strengths and weaknesses.

Remember that AI is not entirely objective and not always accurate. The AI's algorithm collects and aggregates what it finds on the Internet, reflecting the prevailing information it finds. A rosy public relations campaign or a negative news cycle can bias the information the AI gathers, so do some additional research on the Internet to ensure AI-generated results match reality. You should also be cautious about providing data about your organization to an AI application hosted by a third party. Never assume that a third-party AI application is secure because it will be harvesting any information or data you put into it.

When you are finished weighing your and your competitor's strengths and limitations, you should identify the main strength that sets your team, company, or organization apart from the competition. You can use this strength as an overall main point for your Qualifications section.

Drafting the Qualifications Section

Like the other major sections in the proposal, the content of the Qualifications section should be organized into an opening, body, and closing.

Opening

The opening of the Qualifications section, usually a short paragraph or two, identifies the purpose of the section and states a main claim (i.e., the main strength) about the qualifications of your team, company, or organization. In the opening, you want to claim in some fashion that your team, company, or organization is uniquely qualified or best qualified for the project because you possess a particular strength that sets you apart or above your competitors.

At the end of this opening, you might forecast the structure of the Qualifications section. For example, a forecasting statement might say something like, "In this section, we will first describe our management team and workforce. Then, we will explain our mission and history and detail our facilities. Finally, we will describe recent projects we have completed that are similar to the work you want done."

Body

The body of a Qualifications section tends to break down into three areas:

Description of Personnel—biographies of management, demographics of the labor force, special training of employees, security clearances, and certifications.

Description of Company/Organization—corporate/organizational history, mission statement, business or research philosophy, facilities, equipment, patents or proprietary procedures, security and privacy procedures, and quality control procedures.

Experience of Company/Organization—past successful projects, experience of personnel in the industry or field, and successful similar cases to the proposed project.

Description of Personnel

The most important part of the Qualifications section is your description of your company's or organization's personnel. Your readers want to do more than check whether you and your employees can do the job. They want to get to know the people who will do the work. Descriptions of the personnel are typically divided into three areas:

Management—introduces the project leaders, supervisors, and key employees who will occupy specific leadership roles on the project.

Labor—describes members of the workforce who will provide the service, do the lab work, or manufacture the product.

Support Staff—includes assistants, bookkeepers, secretaries, repair technicians, and other employees who will support management and labor.

Your description of the management team should include brief biographies of the managers who will lead the project. Each leader's biography will run one to three paragraphs explaining why this person is an essential leadership team member.

Your management biographies need to provide enough detail to familiarize the readers with your project leaders; however, avoid overwhelming the readers with each manager's life history. In most cases, the project leaders' resumes will be included in an appendix, so provide just enough biographical information to demonstrate that each member of the management team is experienced, properly trained, and motivated to succeed. Also, for each manager, you might want to mention a personal attribute that gives insight into their personality. To give each manager a personality, you can describe their leadership style or mention a personal interest (e.g., travel, volunteer work, special passion, or hobby).

When describing labor and support staff, you want to give the readers an overall feel for the kinds of employees your company or organization tends to hire. You don't have room to provide a biography of each employee. Still, you can offer an overall description of your employees' training, prior experiences with similar projects, and their commitment to quality. Your description of your labor and support staff should give the readers a general sense of the kinds of people who work at your company or organization.

Overall, the description of your management, labor, and support staff should demonstrate to the readers that your team, company, or organization members are qualified to take on the proposed project. Where possible, highlight aspects of your management team and employees that make them uniquely or best qualified to handle the project.

Proposal writers often debate whether to include an organizational chart that illustrates the management hierarchy for the project. Organizational charts take up a significant amount of space in a proposal, and they can be a momentum killer for the readers, so you should use one with care and only when needed.

Organizational charts are best used when the project is complex and numerous levels of managers are involved, as shown in Figure 7.2. That way, the readers can see who answers to whom in the project's management structure. The readers can also see how your company's or organization's hierarchy aligns with theirs, so they can figure out who is coordinating with whom.

Overall, though, we recommend not including organizational charts for smaller projects. After all, your readers have limited time and patience. If an organizational chart simply shows the apparent connections among managers and employees at your company, you probably don't need it.

Figure 7.2: An Organizational Chart

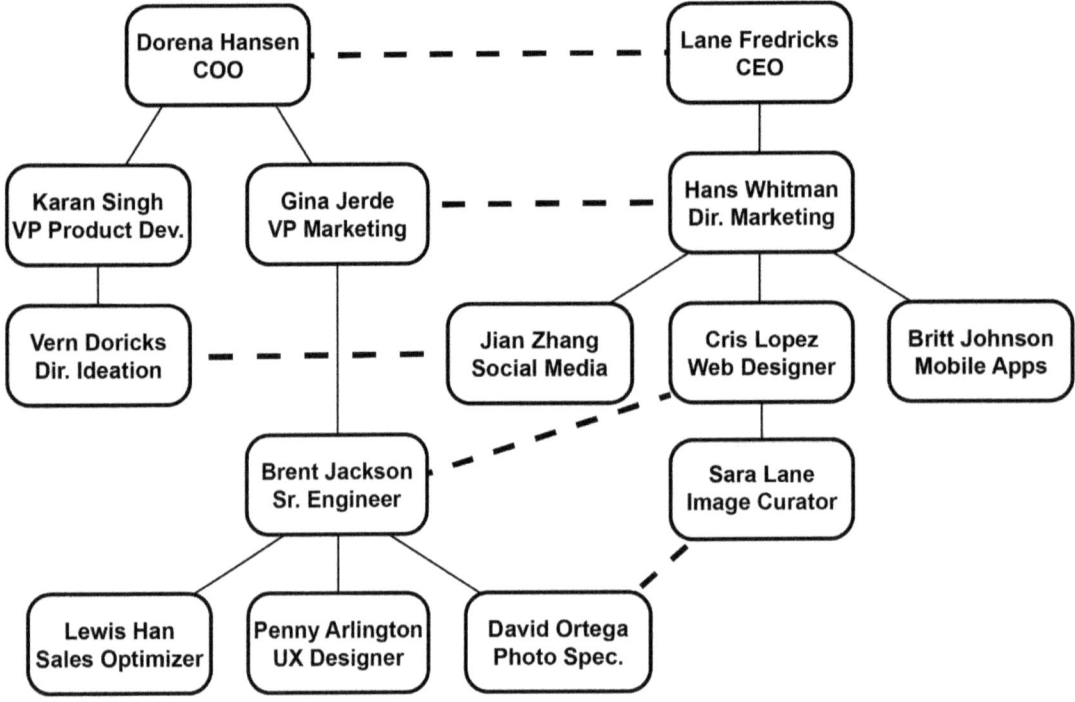

Description of the Company/Organization

As the writer of the proposal, you are familiar with your company or organization, but your readers won't know much of its history, mission, or operations. Even when two companies have long-standing business relationships, a description of your company or organization may be needed because relationships change over time. In other words, two companies that began doing business decades ago don't have the same managers and employees as in the past. Their missions have evolved, and their facilities and equipment have changed over time.

In your description of the company or organization, you want to offer the readers an overview of your operations. The following elements can be included in a description of the company:

Corporate or Organizational History—describes when the company or organization was founded, by whom, and for what purpose. A history often describes the evolution of the company or organization into its present form.

Mission or Vision Statement—identifies your company or organization's goals and principles of operation.

Corporate Philosophy—explains the company's or organization's approach, management style, and the standards by which it does business.

Facilities and Equipment—describes resources, manufacturing capabilities, and machinery.

Quality-Control Measures—highlights compliance with specific quality-control guidelines or international standards like IS0-9000 and ISO-14000.

Your description of the company or organization should use details, preferably quantifiable details, that describe its operations. Be specific about dates in your company's or organization's history. Use numbers to describe the size and output of your facilities. Tell them what kinds of equipment and how many people you will commit to the project. These quantifiable details will add a strong sense of realism to your Qualifications section.

Experience of the Company/Organization

Proposal writers often round out a Qualifications section by listing successful past projects. You want to show a track record of success with similar projects. At the very least, you should show that your company or organization has experience in the industry or the research area. A Qualifications section might strategically list a few similar projects completed recently. In other cases, you may list the companies or funding sources with whom your team, company, or organization has worked before.

Again, numerical details are important in descriptions of your company's or organization's experiences. Where possible, you should use numbers to support your claims about prior projects, especially if you can provide quantifiable evidence that aligns with your readers' key performance indicators (KPI). These quantifiable details will add a sense of realism and help you align yourself with your readers' goals.

Closing

Qualifications sections often don't need a closing, especially in smaller proposals. But, if the body of the Qualifications section still feels open-ended or unfinished, you might include a brief closing paragraph that restates a new version of the claim you made in this section's opening paragraph (i.e., what makes you different). Keep the closing brief, maybe two to three sentences.

Creating a Persona

When writing the Qualifications section, you should be conscious of the persona or image you want this section (and the whole proposal) to project about your company or organization. To identify this persona, you might consult with your company's or organization's public relations team and materials. Specifically, you might incorporate a brand, slogan, or keywords that your company's or organization's public relations materials project to the public.

If your company or organization does not have a defined image or persona, mapping can help you create one for your proposal (Figure 7.3).

Figure 7.3: Mapping a Company's or Organization's Persona

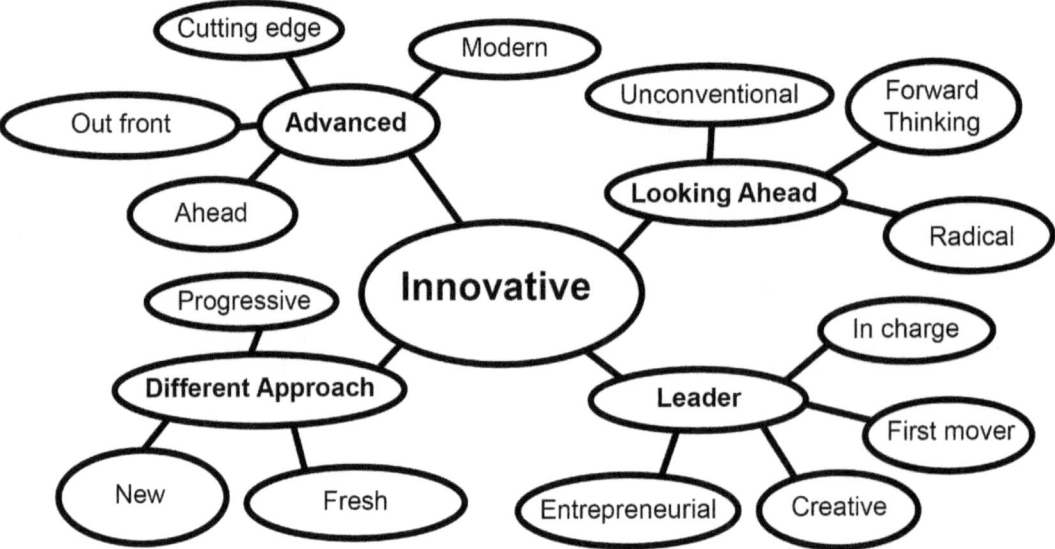

The following steps will help you develop a persona map for your proposal:

1. Review the Strengths and Limitations Worksheet and your analysis of what makes you different and attractive.

2. Find a keyword or phrase that characterizes your company or organization. Is your organization "innovative," "aggressive," or "reliable"? Is it "technologically advanced" or "highly experienced"?

3. Write your keyword or phrase down on a whiteboard, screen, or a sheet of paper and circle it. Start coming up with synonyms and phrases that reflect the qualities indicated by this term. Circle them and connect them to the word or phrase you started with.

4. Mapping out further, find more words and phrases that reflect the persona of your company or organization.

When you are finished mapping out the persona, you will have created a set of words you can now weave into your proposal's Qualifications section. These keywords and phrases in the Qualifications section will help you create a persona by developing what professional writers call a *theme* in your writing. If you use these words and phrases periodically in this section, your readers will begin to hear that theme as they read silently.

Of course, a persona is most effective when the keywords reflect your company's or organization's true persona. If your company is rather traditional or conventional, your attempts to weave a "dynamic" persona into your proposal probably won't work. Likewise, if the people at your company are free-spirited or avant-garde, attempts to project a buttoned-down traditional image will likely backfire. After all, the persona you choose eventually needs to measure up to reality.

To Boilerplate or Not to Boilerplate

Before long, you will run into the word *boilerplate* when discussing proposals and grants. Boilerplates are standard descriptions of a company or organization, often copied and pasted into proposals as the Qualifications sections, sometimes with minimal revision.

The problem with boilerplates is that these generic one-size-fits-all descriptions can be a hodgepodge of materials that have been collected about the company or organization. They regularly contain outdated or irrelevant information to the proposed project. The boilerplate has sometimes ballooned into a multi-page monstrosity because it has become a catchall of materials from multiple sources. Boilerplate Qualifications sections also tend to project an impersonal tone that probably won't match the tone in the rest of the proposal.

Should you avoid using that old boilerplate? Yes, but that doesn't mean you need to start writing every Qualifications section from scratch. Professional proposal writers and grant writers collect assorted files of qualifications-related information, such as short biographies of key personnel, a mission statement, a description of facilities, and so on. These preexisting files can help you generate the content for a Qualifications section unique to your proposal.

You should then revise and rewrite these disparate parts into a cohesive argument that aligns with the project you are proposing. Trim down the Qualifications section to only need-to-know information, stripping away any content about personnel, your company, or your organization that isn't relevant to the project. Make sure the tone and voice are the same as the rest of the proposal.

You might find an AI application helpful for doing this kind of work. You can insert your company's boilerplate materials and tell the AI what persona you want it to develop. More than likely, the AI will still include more information than necessary for your proposal, but it will help you organize the existing materials and make them sound consistent. Then, you can do your own revisions. Again, be careful about copy-pasting proprietary or sensitive information into third-party AI applications. You should assume that the AI is harvesting any information you put into it.

When deadlines approach, you may be tempted to use the old boilerplate with minimal adjustments. Fight that temptation. The few hours spent customizing the Qualifications section to the project will be rewarded with a leaner, more focused argument that stresses your company or organization's unique strengths and qualities.

Chapter Summary and Looking Ahead

The Qualifications section is very important to the readers. You should remember that proposals do more than simply offer plans and budgets. They also build relationships between people and between organizations. They establish organizational *ethos* or credibility. Keep in mind that the readers are not merely looking at the proposal's facts. They are deciding whether they trust you, your company, or your organization.

In the end, people hire people, not plans or budgets, so it is critical that you show your readers how your team, company, or organization is experienced, reliable, and honest, as well as responsive to their needs.

Try This Out!

1. Find a business proposal on the Internet that includes a stand-alone Qualifications section. Consider the following questions:
    a. Does the section effectively argue that the bidding company is uniquely qualified?
    b. Does the section have a clear opening, body, and closing?
    c. What topics did the authors choose to include in their Qualifications section?
    d. What information did they leave out? Why?
    e. Write an analysis to your instructor in which you critique the content and organization of this Qualifications section. In your analysis, point out any strengths and make suggestions for improvements.

2. With a team, write a one-page Qualifications section for your university or a company where one of you works. Identify what makes your university or this company unique or different from its competitors. Consider the following:
    a. What topics (e.g., mission, history, experience, management, employees, or facilities) should you mention in a Qualification section?
    b. What are some topics you might not mention? Why?
    c. Also, how might you turn your university's or company's differences into strengths?

3. Study the Qualifications section from a real proposal. What persona, if any, is woven into this section? Underline some of the keywords and phrases used in the Qualifications section. Do they create a persona that the writers are trying to reinforce? If not, use mapping to develop a new persona for this Qualifications section with concept mapping.

4. Create a concept map that describes the persona of your university or company. If you had one positive word to describe your university or the company, what would it be? Put this word in the middle of a whiteboard, screen, or sheet of paper, and map out the persona associated with this word. Then, weave this persona into the Qualifications section you wrote for Exercise 2.

5. In your class, interview one of your classmates or colleagues. Write a oneparagraph biography of this person in which you discuss their education, experience, and special abilities. Also, add a statement that personalizes the biography.

Case Study: The Carbon Neutral Campus— Who Can Do the Project?

The Carbon Neutral Campus team felt good about what they had accomplished so far. They had identified and described the climate crisis issues that Durango University was facing, and they had developed a project plan that described positive steps they could take toward solving that problem.

Now, they needed to write a Qualifications section for the grant proposal. The university had plenty of administrators who could lead the project. Plus, several faculty members on campus were specialists in environmental issues and alternative energy research. One of their team members, George Tillman, was a professor of environmental engineering specializing in renewable energy.

The grant-writing team decided to let Anne and George write this part of the grant because they were in the best position to identify people who could lead the project. Anne and George set up a video conference to do some brainstorming.

"George, we need someone like you to be our project leader," said Anne. "You have experience in these kinds of issues."

"Sure, I have some experience with environmental engineering," George replied, "but I don't know anything about urban planning or getting larger groups of people together. I first heard the word 'charrette' when Calvin said it last week."

Anne said, "I've thought about that already, so I started looking for someone at Durango who specializes in urban planning and environmental issues. Unfortunately, we don't have anyone like that who does that kind of work. Usually, we bid out these large-scale projects to firms specializing in campus planning. Several consulting companies do this kind of work."

"Does Calvin remember the name of that urban planning firm that ran the charrette in Albuquerque?"

"Yeah, I was wondering that too," said Anne. "Calvin sent me a link to their website, and I called them. The firm that ran the charrette was Summers & Mondragon, located in Santa Fe, New Mexico."

"That's not too far from us," said George. "They're only a quick flight or four-hour drive from here."

"I talked with one of their partners, Diane Smith, who specializes in environmental urban planning. She told me they had never worked on designs for a university campus. So, before doing any design work, they would need to research our campus's needs. But she felt they could at least run a charrette for us."

"Do we need to bid out this kind of project?" asked George.

"Not this part of the project. Running the charrette and developing the first set of designs is a consulting job, so we can hire a firm without putting out the contract for bid. But, when we move beyond that point to hire a contractor to do the work, we will need to bid out the project."

"So, you're saying we should propose hiring Summers & Mondragon to facilitate the charrette?"

"Yes," answered Anne, "and I want to put you and me down as the co-project leaders who will lead the planning process."

"Who else needs to be named project leaders?" asked George.

"I think we need to recruit one of our librarians to be on the planning team because setting up an online library of sources and managing the website for the charrette will be a big part of the project."

George thought for a moment and then said, "I know Gina Sanders over there. She has worked with the Department of Environmental Engineering to collect information on several projects. She's knowledgeable about environmental issues."

"Sounds great," replied Anne. "We'll ask her if she's interested."

"Now comes the hard part," said George. "Why us? Why should the Tempest Foundation provide funding for a project like this one? Why Durango University? Surely, we're not the only people considering converting our campus to renewable energy sources. Other universities may ask the Tempest Foundation to fund a similar project. What makes us different?"

Anne thought a moment. "That's a good question. Certainly, more prominent universities are out there looking for this kind of funding. We need to figure out what sets us apart from our competitors for funding."

"Well, our location is one difference. We get lots of sun and wind here. We can also tap into the geothermal energy that's available in a mountainous area like ours," said George. "That gives us several proven options for renewable energy."

Anne added, "Another thing that makes us different is the size of our campus. We're not big, so we would offer a good test site for this concept. If the conversion to renewable energy works here, other campuses could use our experience to guide their conversions to non-carbon energy sources."

"We also have a solid Environmental Engineering department. It was one of the first in the southwestern U.S. We've received grants from the National Science Foundation to research geothermal energy and other kinds of alternative energy."

"Yes. We can build that experience into the Qualifications section."

"Most of all," said George, "I think we have a strong commitment to making this transformation happen. We all want to make this work from the university president down to our students. It's not just an academic exercise. We actually want to do this. Mentioning our commitment might set us apart."

For the rest of the hour, George and Anne roughed out a draft of the Qualifications section for the grant proposal. They wrote bios for themselves and the rest of the Carbon Neutral Campus Team. They left a space that Diane Smith from Summers & Mondragon could fill with their own bios and corporate background information. They also sent an email to Gina Sanders, the librarian they wanted to recruit.

They used an AI application to summarize information from the university's website to fill out the Qualifications section's descriptions of the university and its accomplishments.

That afternoon, Anne called Diane Smith at Summers & Mondragon to see if their consulting firm would be interested in discussing the project. Diane said she had already mentioned the project to her colleagues, and they were interested.

The next day, Gina Sanders, the university librarian, got back to them and said she was interested in joining the project.

Anne and George asked Diane and Gina to send them brief biographies and backgrounders on their organization that could summarized and inserted into the grant.

George finished a rough draft of the Qualifications section later that week and sent it to the rest of the grant writing team for comments.

If you would like to see how the Qualifications section of the Carbon Neutral Campus proposal turned out, please go to Appendix A.

8 WRITING A CONCLUSION WITH COSTS AND BENEFITS

Finishing on a High Note

Your readers probably won't read your proposal cover to cover, but they almost always carefully read the conclusion of a proposal or grant. So, you want this final section to present the best case for saying yes to your proposal. The first time through, your readers will usually read the introduction and then skim the Background, Project Plan, and Qualifications sections. But they will almost always read the conclusion closely. Why? Your readers know that the conclusion is where you will make your best case for accepting the proposal. Here is also where you will tell them how much the project will cost and highlight the benefits of accepting your proposal.

Your readers will ask two important questions before making any decision: "How much will this project cost?" and "What do we get if we say yes?" A well-written conclusion answers these questions in a straightforward and persuasive way. It should also bring the readers around to the beginning of the proposal. In your introduction, you told them what you were going to tell them. In the body of the proposal, you told them. Now, in the conclusion, you are going to summarize the whole argument and tell them what you told them.

Conclusions tend to be brief in proposals. Usually, the more you say in a conclusion, the less your readers will remember. So, you want to boil the proposal's argument down to its costs and benefits, allowing the readers to make a simple costs-benefits analysis. If they decide the project's benefits are worth the costs, your proposal has a good chance of being accepted.

Identifying the Benefits

In proposals, as well as in life, the best way to persuade people is to stress the benefits of saying yes to your ideas. Bottom line: What do they get if they say yes? There are three types of benefits you can mention in your conclusion:

> **Hard Benefits**—measurable deliverables, such as products, services, increased revenues, more customers, lower costs, improved outcomes, and better results
>
> **Soft Benefits**—intangible advantages like quality, loyalty, service, satisfaction, goodwill, flexibility, and happiness
>
> **Value Benefits**—common values and standards your company or organization shares with your customers, clients, or the funding source.

Most of your proposal's conclusion will usually summarize these hard, soft, and value benefits. Fortunately, in one way or another, you have already identified all these benefits in the body of the proposal, so now you need to bring them together in one place.

Hard Benefits: Deliverables

Hard benefits are the project deliverables that the readers will see, touch, or interact with. Hard benefits tend to be quantifiable, meaning they can be counted or measured. They work on two levels:

- The customers receive *direct benefits* from possessing the deliverables identified in the Project Plan section (e.g., a product, service, factory, prototype, public relations campaign, report).
- The customers will also receive *consequent benefits* that accompany these deliverables (more profit, more efficient employees, better public relations with their customers, a solution to a problem, or a plan for taking advantage of an opportunity).

For example, the direct benefits of an implementation proposal are tangible things like a new product, building, or public relations campaign. These benefits (deliverables) are quantifiable objects you will deliver to the readers as the project is happening or when it's completed.

You can also highlight the consequent benefits of receiving these direct benefits to the readers. A new factory, for instance, should make their employees more efficient or allow the company to expand their manufacturing capabilities. A new product might mean more market share that brings in more revenue. A public relations campaign might result in a higher corporate profile or better relations with the community. These are all consequent benefits that happen because of the direct benefits.

To identify your proposal's direct and consequent hard benefits, review your Project Plan section and list all the deliverables you promised your readers. Using a Benefits Worksheet, like the one shown in Figure 8.1, write those deliverables in the "Deliverables" column. Then, in the second "Added Benefits" column, write down all the additional advantages the readers will receive with each of these deliverables. As much as possible, view these deliverables from your customer's perspective. Ask yourself how each deliverable will make their situation better than it was before.

Soft Benefits: Qualities

Soft benefits are the advantages of working with your company or organization, including intangible benefits like trust, efficiency, satisfaction, and confidence. These soft benefits usually cannot be held, seen, or even measured. Nevertheless, they are important benefits that you can offer to your readers.

To find these soft benefits, look back at your Qualifications section and list out the strengths of your company or organization. Pay special attention to how you set your company or organization apart from its competitors. Remember: What makes you different makes you attractive.

List these soft benefits in the third column, "Strengths," of the Benefits Worksheet (Figure 8.1). Then, in the fourth column, "Added Benefits," write down the advantages these soft benefits will bring your readers.

Of course, you can make any claims you want about soft benefits like high quality or customer satisfaction, but the best soft benefits are ones that highlight special qualities unique to your company or organization. For example, if your company has a reputation for high-quality work, then stressing quality in the conclusion would be a good idea. However, if your company is not a quality leader in your field, your attempts to sell your readers on quality will come off a bit hollow. You're better off focusing on your company's or organization's existing qualities, perhaps lower cost or a quick turnaround, rather than selling them on things that aren't true.

Figure 8.1: Benefits Worksheet

Hard Benefits		Soft Benefits	
Deliverables (Direct Benefits)	**Added Benefits (Consequent Benefits)**	**Strengths**	**Added Benefits (Consequent Benefits)**

Value Benefits

Value Benefits: Shared Worldview

Value benefits refer to the shared values held by your customers or clients and your team or company. Review the Reader Analysis Worksheet you filled out in Figure 3.3 of Chapter 3. In that worksheet, you wrote down your primary readers' needs, values, attitudes, and emotions concerning the project. In the Values column of the worksheet, you identified some qualities that the readers value in themselves. It's relatively safe to assume they value these qualities in the people they hire to do a project for them.

List out these value benefits in the Benefits Worksheet (Figure 8.1). As you draft your conclusion and revise your proposal, you will want to weave in the value benefits to signal your readers that your company or organization shares their values.

These hard, soft, and value benefits will have been brought up somewhere earlier in your proposal, so you're not telling the readers anything new. You're summarizing all these benefits at the end of the proposal, so the readers know what they get for their money. That way, your readers will have all these benefits in mind as they decide whether to say yes.

In most proposals, the only *new* information in the conclusion will be a summary of the costs of the project, product, or service. As you might guess, revealing the costs is a key moment in any proposal because your readers will be looking for the price tag. The bottom-line cost can also be a bitter pill for your readers because you are telling them they need to trade something, usually money, to get something they need.

Fortunately, the costs can be strategically positioned in the conclusion, so they are sweetened by the benefits. By considering the costs and benefits together, your readers will understand precisely what their money will purchase.

Drafting the Conclusion

The conclusion is typically shorter than the other sections of a proposal. Proposal writers often use a gymnastics phrase, "stick the landing," to remind themselves that no matter how complicated the body of the proposal, the conclusion needs to land squarely on its feet—with a smile.

Your conclusion should be positive and forward-looking. You shouldn't rehash the problem or try to scare the readers into saying yes. Rehashing the problem may put your readers on the defensive and leave them in a bad mood. No one likes to be pressured or threatened into agreeing with a proposal (e.g., "Accept our plan or bad things will happen."). Instead, your readers want to feel as though accepting the proposal will improve their lives and make things better. In most situations, your proposal's conclusion should be positive in content and tone.

The conclusion of the proposal can be built around five specific moves. These moves are intended to stress the plan's benefits while stating its costs.

Move 1: Make an obvious transition from the body of the proposal.

Move 2: State the costs of the plan.

Move 3: Summarize the benefits of the plan.

Move 4: Look to the future.

Move 5: Thank them and identify the next step.

You do not need to make all these moves in the conclusion, and they don't need to be made in this order. A combination of some or all these moves will round out the proposal in a strong way. Let's look at them closely.

Move 1: Make an Obvious Transition from the Body of the Proposal

The moment when your proposal transitions from the body to the conclusion should be obvious to the readers. Keep in mind that they've been skimming or reading for a while, so you want to wake them up and catch their attention again. Your proposal has already given them numerous details to sort out and mull over. An obvious transition will signal that they are in the home stretch.

There are three main ways to make an obvious transition into the conclusion:

- Use a heading to cue the conclusion while signaling the readers that you are wrapping up. A heading like "Concluding Remarks," "Our Recommendations," or "What Are the Advantages of our Plan?" will usually prompt them to start paying closer attention.
- Use a transition, such as "In summary..." "To conclude...," or "To wrap up...." A transitional sentence might say something like, "We will now conclude by reviewing the costs and benefits of our...."
- State or restate your proposal's main point. Your main point sentence will signal that you are bringing the proposal to a close.

The transition into the conclusion is a critical moment in any proposal or grant, so it should be handled with care. Once the conclusion has been signaled, you usually have a few paragraphs, at most a page, to state the costs and explain the benefits. If you go longer, your readers will become frustrated, restless, or, worst of all, bored.

Move 2: State the Costs of the Plan

The cost is the bottom line of any proposal. Unfortunately, finding out how much a project will cost is typically not a high point for your readers—unless you are asking for significantly less money than they expected! So, you should do your best to present the costs in the best positive light.

Presenting the Costs

There are a few strategies for presenting costs in a constructive way. In smaller proposals, the project's costs can be concisely itemized in a short table with a few explanatory comments. The table in Figure 8.2, for example, shows how major budget items can be concisely summarized for the readers.

Figure 8.2: A Table that Summarizes the Major Budget Items

The costs of the project are as follows:	
Development of prototype	$105,390
Retooling manufacturing facility	$890,600
Retraining labor and staff	$192,200
Promotional campaign	$120,000
Contingency	$50,000
TOTAL	$1,358,190

In larger proposals, writers sometimes include a stand-alone Budget section sandwiched between the Qualifications section and the proposal's conclusion. In these Budget sections, the readers typically find an itemized table of costs and a detailed rationale for major and minor expenses. These sections can often run on for several pages.

We recommend putting the full itemized budget in an appendix, not inside the body of the proposal. Budgets are momentum killers in proposals. So, if you place a full budget, tables and all, inside the body of your proposal, you will sidetrack the overall story you are trying to tell. By the time the readers finish grinding through a few pages of budgetary items, dollar figures, formulas, and explanations, they will have forgotten the core argument of your proposal. If they are tired of reading, they might not even get to the benefits you are promising them!

Offering a Synopsis of the Costs

There is a better way to handle the costs in a proposal. If possible, you should include a brief overview of the project's costs in the conclusion of the proposal. Then, refer them to the itemized budget and a detailed rationale in an appendix. For example, you can write, "As shown in our budget in Appendix B, we estimate the Goodman Restoration Project will cost $4,560,300." This straightforward statement of the bottom-line cost is sufficient for even the largest projects. You are providing the readers with a bottom-line figure while directing them to a place where they can study the costs in greater depth.

The advantage of this approach is that the readers avoid slogging through a detailed discussion of costs just before they decide to put your proposal in the "maybe" or "no" pile. If the readers want a comprehensive discussion of the costs, they can turn to the appendix, where the figures are handled in depth. (Chapter 9 will discuss how to write stand-alone budgets for an appendix or a Budget section.)

If your proposal's conclusion needs more than one sentence about costs, you might include a small table that breaks down the budget into larger parts, much like the table shown in Figure 8.2. An overview budget table summarizes the basic costs of the budget without breaking down those figures into itemized dollars and cents. Keep things simple!

Stating the Costs in a Straightforward Way

State the costs of your project directly and unapologetically. Sometimes, writers feel a strange urge to go weak in the knees when stating how much the project will cost. But attempts to soften, defend, or put a positive spin on costs tend to backfire.

Instead, when you state how much the project will cost, just tell the readers the figure in a straightforward way. Apologizing, defending, or putting a phony-sounding sales spin on the costs rarely has a positive effect on the readers. Meanwhile, your customers or clients may see your apologizing or defensiveness as an opening to negotiate a lower price.

Remember that a proposal is a *de facto* contract until it is replaced by a real contract. Once the proposal is accepted, it is usually considered a temporary contract until a formal contract can be generated and delivered. For smaller projects, the proposal may be the only "contract" that is signed.

So, if you promise that you can complete the project for a specific amount of money and your customer or client accepts, then assume you are obligated to charge only that amount. Therefore, the costs in a formal proposal should be stated in a straightforward way, and they need to be defensible in a court of law, if necessary.

Move 3: Summarize the Benefits of the Plan

Most of your proposal's conclusion will be devoted to summarizing the benefits of saying yes to the proposal. A good strategy is to offer a bulleted or numbered list of hard benefits that the customer or client receives if they say yes.

You should limit the number of benefits to five to seven major items. You might be tempted to generate a long list of all the benefits, both large and small, but you want to focus on the major benefits that will be most attractive to the readers. An attempt to list every possible benefit might cause the major benefits to become lost in a fog of less-important benefits.

A good technique is to weave the soft benefits into your discussion of the hard benefits. For example, instead of only telling the readers they will be purchasing a new latte machine for their café, you can also point out that they will be gaining "added benefits" like bolstering the excellent service the café is known for and cultivating stronger customer loyalty.

An excellent way to wrap up the hard and soft benefits discussion is to include a small paragraph highlighting one or two value benefits. If your customers or clients stress efficiency or trustworthiness in their marketing documents, here would be a good place to echo those value benefits back to them.

You might try using an artificial intelligence (AI) application to generate a list of deliverables, soft benefits, and value benefits for you. You can copy your Project Plan section into an AI application and ask it to identify your deliverables and soft benefits. Then, copy your Qualifications section into the AI and ask it to identify some additional soft benefits. Finally, the AI application should be told to identify any value benefits on the customer, client, or funding source's website and social media pages.

As always, if you are using a third-party AI, make sure you are not including proprietary or personal information. The AI will almost surely harvest that information.

Move 4: Look to the Future

Whether you are solving a problem for them or helping them take advantage of an opportunity, the bottom-line promise you are making to the readers is that their future will be better if they say yes to your proposal or grant application. For this reason, many writers like to include a "Look to the Future" paragraph that visualizes that better future for the readers. For example, if you are proposing a new manufacturing facility, you can describe the company's employees working efficiently in this new state-of-the-art facility. Or, if your grant is proposing a new summer jazz festival for your town, you can briefly describe people in the community swaying to the music on a summer day.

A look to the future in the conclusion is usually only a brief paragraph. It aims to *show* the readers that your project, product, or service will create the future they desire.

Move 5: Thank Them and Identify the Next Step

A good way to put the final touches on a proposal or grant is to thank the readers for considering it. Give them contact information if they have questions. Then, identify your reader's next step.

Their next step is what you want them to do immediately after reviewing your proposal. Do you want them to contact you by text, email, or phone? Should they set up a video conference with you? Will you be calling them or scheduling a visit? The readers should finish your proposal with a clear idea of what action is needed if they want to accept it.

Why is a next-step statement a good way to end the proposal? It's fairly common for customers or clients to look over a proposal with approval, even excitement. Then, they set it aside for the moment because they aren't always sure what needs to happen right now. A next-step statement, sometimes called a "call to action," tells them to contact you (or wait for you to contact them). It identifies the small, concrete steps they need to take right now to put the project into motion. Even million-dollar proposals need a starting place. Sometimes, that starting place is a quick text, email, or phone call to set up a meeting.

Chapter Summary and Looking Ahead

Your proposal's conclusion should explain the costs and benefits of the proposal, and it is often the most important part of the document. Readers will skim some parts of your proposal and read others more closely, but they will almost always read the conclusion. So, you want your conclusion to bring the readers back to the proposal's main point while stressing the project's importance. An effective conclusion amplifies the benefits of saying yes to the plan.

In the next chapter, we will discuss how to write budgets and budget narratives. In the remaining chapters, we will discuss how to develop your proposal's style and design it to be attractive and accessible. When you are finished drafting the conclusion, though, you should feel some satisfaction. After all, you are on the home stretch in the proposal-writing process.

Try This Out!

1. Look at the conclusion of a proposal or grant you found on the Internet or at your workplace and consider the following questions:
    a. Does the proposal make a clear transition from the body to the conclusion?
    b. Where and how are the costs handled in the proposal?
    c. Did the authors present the costs in a straightforward and unapologetic way?
    d. Does the conclusion highlight hard, soft, and value benefits?
    e. Does it offer readers a look into the future?
    f. And finally, does it identify the next step, or trigger, that will put the proposal into motion?
    g. Mark specific sentences where the writers made some or all the five moves typical in a proposal's conclusion. If some of these concluding moves are missing, do you think the writers of the proposal had good reasons for leaving them out?

2. A common problem in most workplaces and campuses is obsolete computers and computer networks. Using a Benefits Worksheet like the one shown in Figure 8.1, list the hard, soft, and value benefits gained if your university or workplace upgraded its computer hardware or software.

3. For a practice or real proposal of your own, write a conclusion for a proposal that includes only the five moves described in this chapter. Pay special attention to the benefits you identified in the body of your proposal. Summarize those benefits in your conclusion.

Case Study: The Carbon Neutral Campus Project—What Are the Costs and Benefits?

When they finished drafting the proposal's larger sections (Introduction, Background, Project Plan, and Qualifications), the Carbon Neutral Campus team was ready to write the conclusion.

Karen and Tim had already been looking through the Project Plan and Qualifications sections to identify any deliverables and other hard benefits promised.

As the others joined the meeting, Karen said, "OK, we should summarize the most important benefits at the end of the proposal so we can show the Tempest Foundation what they are paying for."

As Karen and Tim highlighted the deliverables, George walked to the whiteboard and drew a benefits chart with four columns (Figure 8.3). He turned and said, "OK, when you're finished finding those deliverables, let's put them into this chart.

Figure 8.3: The Benefits Worksheet for the Carbon Neutral Campus Project

Hard Benefits

Deliverables	Added Benefits
Zero-carbon campus strategic plan	Comprehensive local action responding to climate crisis
Urban planning charrette	Community involvement in the process
On-going dialogue about sustainable energy practices on campus	Buy-in and awareness from local community
Steering commitee to guide beginning of Cool Campus Project	Flexible, focused team that will be responsible for taking action
Cool Campus library of sources and a website	Local and worldwide access to information about campus conversions to sustainable energy
Model that other universities could follow	
Regular progress reports to the Tempest Foundation	The Foundation will be kept informed and involved

Soft Benefits

Strengths	Added Benefits
Community-centered approach	Bring us together as a community
Variety of local sustainable energy sources	Options and flexibility to explore solutions
DU campus is a good size for attempt	Changes can be made without the project growing out of control
Strong environmental engineering program at DU	Experience is available on campus for advice and insight
History of successful projects related to environmental issues.	Confidence that project will be completed

Value Benefits

Independence to solve our own energy problems
Leadership in environmental issues
Better future for everyone
Knowledge and wisdom of the community

When Karen and Tim had a good list, they began saying the deliverables out loud while George wrote them down on the whiteboard. They also talked about the added benefits that came with each deliverable. Figure 8.3 shows how they filled out the chart.

They also listed the soft benefits, including some of the non-measurable advantages of partnering with Durango University. They added a few benefits that came with each soft benefit.

When that was finished, they looked closely at the Tempest Foundation's website, paying close attention to the values expressed in the organization's mission statement and descriptions of previous projects. They put these items down in the values benefits area of the table.

The team began roughing out the conclusion together by going through the five concluding moves.

Generating content for the conclusion wasn't too difficult. They wrote down an obvious transition (i.e., "Let's conclude... ") and then restated the purpose and main point of the project. The purpose was, "We are requesting $ XXX,000 from the Tempest Foundation to help us develop a Carbon Neutral Campus Strategic Plan." The main point was, "With this strategic plan in place, we can begin taking positive steps toward addressing the causes of the climate crisis while serving as a model for other college campuses."

They used a bulleted list to itemize the major benefits of their project. They couldn't mention all the deliverables and benefits listed on their worksheet. There were just too many. So, they concentrated on the ones that seemed most important or impactful.

They then brainstormed about a Look to the Future paragraph that would paint a positive image of a sustainable future made possible by renewable energy.

In the final paragraph, they thanked the reviewers and provided Anne's name, phone number, and email address. This personal touch seemed to end the grant proposal on a positive and professional note.

Tim said, "Hey, this looks great, but there's an obvious problem. We don't have an estimate for this project's cost."

"Yeah, that's always the tough part," smiled George. "We still need to work out the budget. If you don't mind, I will sketch out the budget and schedule a meeting with the development officer for the College of Engineering. His name is Bill Vinn, and he's always really helpful."

"While you're doing the budget, I'll start putting the grant proposal together and work on making it sound clear and consistent," said Anne. She turned to Calvin. "Calvin, can you do some document design to make the proposal more readable and attractive?"

"Sure thing." Calvin was happy to do something more design-oriented, leaving the budget numbers and editing to the others.

Tim spoke up. "I'll use an AI to generate some graphics and images to add to the proposal. We need to show as well as tell. A few images, charts, and graphs should help."

At this point, they were tired of working on the grant proposal, but they were also excited about completing the first draft. There was still a good amount of work to be done, but they were getting closer all the time. They agreed to meet a week later with their parts of the project completed.

If you would like to see how the Conclusion of the Carbon Neutral Campus proposal turned out, please go to Appendix A.

9 BUDGETING A PROJECT

Budgets: The Bottom Line Is the Bottom Line

Like it or not, budgets can make or break any proposal or grant. After all, the bottom line *is* the bottom line. Customers, clients, and funding sources will scrutinize your project's budget with the eyes of a hawk. They will challenge the weak points in the budget, and they will try to cut away what they perceive to be fat. The more your readers cut out of your budget, the less flexibility you will have to complete the project. So, you want to be certain that your proposal's budget is sound and defensible.

For this reason, you should find a good accountant or business manager to help you address the budgetary issues in a proposal. Let's be honest. Many people have trouble with their household budgeting. Creating a budget for a project can be vastly more complicated. To avoid these complications, a good accountant or business manager can help you anticipate and avoid some of the pitfalls of budget development. They can also flag costs you may have overlooked or forgotten to include. You should visit an accountant or business manager soon after you have written a rough version of your proposal to identify costs and clarify the costs you already anticipate. If you work for a larger organization or company, you need to contact the accounting department earlier than later.

This chapter is not designed to substitute for a good accountant or business manager. Instead, it will discuss the basics of project budgeting for a proposal or grant. That way, you can sketch a rough budget before meeting with your accountant or business manager. Remember that your proposal serves as a *de facto* contract between you and your readers, so you need to ensure that your budget is accurate and acceptable to all parties.

Budget Basics

Like any profession, accountants and business managers have their own vocabulary. Let's start by reviewing some of the basics of budget development and accounting terminology.

Itemized and Nonitemized Budgets

Budgets can either be itemized or nonitemized.

Itemized Budgets—Itemized budgets break down the proposal's expenses into their smallest elements (within reason). An itemized budget shows the readers exactly how much money will be spent on each part of the project. It usually has little or no gray area or fungible places. You need to account for every dollar in the project. Itemized budgets can sometimes run several pages in a proposal.

Nonitemized Budgets—Nonitemized budgets, on the other hand, tend to be less exact. They are used when the readers do not need to know where each dollar will be spent. Instead, they receive a rough breakdown of expenses to show them how you came up with the figures in the budget. Nonitemized budgets are shorter than itemized budgets because they consolidate smaller costs into larger categories. In business proposals, the budget given to the customers or clients is often nonitemized because the calculation of costs is considered proprietary information. For smaller grants, a few government foundations, like the National Science Foundation (NSF) and National Institutes of Health (NIH), allow brief nonitemized budgets called modular budgets.

For a few reasons, you should develop an itemized budget whether you intend to include it in the proposal or not:

- *An itemized budget helps you control your costs.* This more detailed account of costs will allow you to track where the money is being spent, even if your readers would be satisfied with a nonitemized budget in the proposal. As the saying goes, the devil is in the details, and that's especially true with budgets. Itemization will help you find hidden costs that would bleed money from your budget.
- *Clients and funding sources often ask you to explain how the budget was calculated.* For example, if a line on your budget mentions a 100-page training packet that costs $190.00 each, your customers may want to know exactly how you came up with that figure. (After all, copying each packet would only cost them $10.00 if they ran it off on their copier.) An itemized budget would help you point out in exact amounts how production expenses, such as labor and overhead, would add to the costs of the packet.

You can convert an itemized budget into a nonitemized or modular budget, so it is best to create the itemized budget first.

Fixed and Flexible Budgets

Budgets can also be *fixed* or *flexible*.

Fixed Budgets—With a fixed budget, you promise to provide the readers with a particular product or service for a set price. That bottom-line price does not change in fixed budgets, even if production costs rise or fall during the project. For example, when using a fixed budget, if the costs of the materials used to make a product suddenly rise, you cannot return to the customer, client, or funding source and expect them to increase your budget to cover those additional expenses. On the other hand, if the costs of your materials drop, you can usually keep those savings. A fixed budget means the project's total price will not increase or decrease even if inflation rises or supply chain issues increase your costs.

Flexible Budgets—Sometimes, the budget needs to be flexible, especially with internal projects within a company or organization. A flexible budget is usually adjusted monthly, quarterly, or yearly to reflect changes in project costs. At the end of each reporting period, the projected costs in the budget are compared to actual costs, and adjustments are made for the next reporting period. A flexible budget allows a company to regulate the costs of a project, modifying the budget to suit changes in objectives or the economy. With external proposals, flexible budgets tend to work only when there is a close relationship between a supplier and buyer. In these situations, the proposal and its budget are updated periodically to address any significant changes in costs.

If you are uncertain about whether your proposal's budget will be fixed or flexible, you should talk to your accountant and your readers about whether a fixed or flexible budget is appropriate for your proposal.

Fixed, Variable, and Semi-Variable Costs

An accountant will often begin a budget discussion by helping you identify the fixed, variable, and semi-variable costs of the project. Even with a fixed budget, some expenses will fluctuate during the project. By identifying which costs are fixed, variable, and semi-variable, the accountant can anticipate the behavior of these costs and estimate the appropriate figures for the budget. To save time (and money), you should break down your costs into three categories before meeting with your accountant.

Fixed Costs—Fixed costs are expenses that will remain constant throughout a project. These costs might include the rent for facilities or equipment, yearly depreciation of facilities or equipment, and management's salaries and benefits. Fixed costs tend not to change with increases or decreases in production. In other words, even if your company increases output by 15 percent, the fixed costs will generally remain the same.

Variable Costs—Variable costs change proportionally with increases or decreases in production. For example, if your company needs to increase production by 15 percent, then variable costs like labor, materials, and energy will also increase by about 15 percent. An increase in production, after all, means you will need to buy more materials, hire more labor, and pay more energy costs.

Semi-Variable Costs—These expenses fluctuate indirectly with production. For example, if you need to increase output temporarily by 15 percent to complete a project, you may need to pay your current hourly employees overtime at 1.5 times their regular rate. Also, if production increases by 15 percent, the costs to service and repair your equipment may increase only 5 percent. Higher production volume, however, often means lower prices on materials, lowering costs in some cases.

The difference between variable and semi-variable costs is how they fluctuate with production. Variable costs rise and fall in step with production, while semi-variable costs rise and fall in ways that are not in step with production.

For the most part, don't lose any sleep over this three-part division of costs. The division into fixed, variable, and semi-variable costs is mostly a convenient way for accountants to anticipate fluctuations that occur in new projects. These categories provide a starting point to identify how costs might fluctuate when the project is underway.

Developing a Budget

Some customers, clients, and funding sources, especially the federal government, will tell you exactly how they want you to break down project costs. In these cases, you should follow their guidelines as closely as possible. If they do not provide any guidelines on the budget, you can start the budgeting process by using a Budget Worksheet like the one shown in Figures 9.1 (For-Profit) or Figure 9.2 (Non-Profit). These worksheets are generic, so you will need to adjust them to your company's or organization's specific needs.

The worksheets divide your costs into the following categories:

- Management, principal investigators, and salaried labor
- Direct labor or staff
- Indirect labor
- Facilities and equipment
- Direct materials
- Indirect materials
- Travel
- Communication
- Profit (business proposals only)
- Facilities and administrative (F&A) (grant proposals only)
- Cost sharing (grant proposals only)

Depending on whether you will include an itemized or nonitemized budget in your proposal, some of these categories may not be represented in the budget you submit with the proposal. Nevertheless, all costs for a project should be assigned to a category in your itemized budget. Let's look at these categories individually.

Management, Principal Investigators, and Salaried Labor

As a fixed cost, the rates for management, principal investigators, and salaried labor are calculated according to the percentage of their time they will devote to the proposed project (Figure 9.1 and Figure 9.2). For example, the project leader, who makes $220,000 annually, will spend

100 percent of her time on the project for 6 months. In the budget, you would identify her by name, write down her total yearly salary, and then multiply by 0.5 because she is spending half her time that year on the project. So, you would write down $110,000 for this manager's salary on the budget sheet. Another manager who makes $150,000 will only devote a quarter of his time to the project over one year. You would multiply his salary by 0.25 (for the quarter time) to come up with $37,500.

How do you calculate how much time each manager will spend on this project per week? After all, managers spend some days on one project and other days on another. You should estimate how much time, on average, a manager will spend on each project. Then, use that average percentage to calculate how much that person's time will cost. Exact figures aren't needed in most cases, and estimates are usually acceptable when budgeting a project.

Other management-related costs are benefits, training, and any outside consultants or specialists your managers or principal investigators will require to do their jobs. Benefits include health insurance, vacation, sick leave, and retirement. Your accountant or business manager can calculate the costs for these benefits. Training may include any special courses or learning tools your project managers would need to prepare for the project.

If your managers or principal investigators need the help of outside consultants or specialists, build the costs of those consultants into this part of the budget. Consultants and specialists are usually compensated on a retainer or contract basis. When preparing the budget, you may need to designate a pool of money from which consultants or specialists will be paid, if needed.

Direct Labor

Direct laborers are the people in your company or organization who are paid hourly. In manufacturing, direct laborers are the people assembling the product. In a store, the direct laborers run the cash registers, stock the shelves, and assist customers at the service counter, among other tasks. In a non-profit organization, the laborers are your staff, lab technicians, or anyone you pay hourly to work on the project.

Direct labor is calculated on an hourly basis. First, you need to estimate how many hours an average hourly employee will spend on the project. Then, multiply those hours by the average amount your hourly employee makes. Of course, employees are paid at different rates, especially in larger companies. Sometimes, it helps to identify these levels in the budget and their respective pay scales. Then, the costs of the employees at each level are calculated separately (Figure 9.1 and Figure 9.2). For example, if ten "Level 3" employees are paid an average of $20.25 per hour, each hour worked by these ten employees will cost $202.50 plus benefits. In a project estimated to take one hundred hours, these employees will cost the customer or client $20,250.00 plus benefits. Meanwhile, twenty "Level 2" employees working at an average of $18.75 plus benefits per hour for one hundred hours will cost $37,500 plus benefits.

Figure 9.1: A Budget Worksheet for a For-Profit Project

Itemized Budget (For Profit Business)

	Estimated Hours or Miles	Rate per Hour or Mile (if applicable)	Estimated Total Cost
1. Management			
a) Project Manager or Principle Investigator			
Benefits			
b) Managers and Supervisors (list out)			
Benefits			
c) Specialists			
Benefits			
d) Consultants			
e) Additional Training			
Total Management Costs			
2. Labor			
a) Direct labor			
b) Indirect labor			
c) Contractors			
d) Training			

e) Labor Overhead
        Fringe Benefits
        Insurance
        Vacation
        Sick Leave
        Pension			

Total Labor Costs			
3. Equipment and Supplies			
a) Computer Hardware, Software. Internet			
b) Tools			
c) Tables, Chairs, Desks			
d) Office Supplies			
e) Depreciation of Equipment			
Total Equipment Costs			

Figure 9.1: A Budget Worksheet for a For-Profit Project (Cont.)

	Estimated Hours or Miles	Rate per Hour or Mile (if applicable)	Estimated Total Cost
4. Facilities and Maintenance			
a) Building Rent			
b) Retooling Facilities			
c) Depreciation of Facilities			
d) Insurance			
e) Internet and Telephone			
f) Water, Electricity, and Gas			
g) Maintenance			
Total Facilities Costs			
5. Materials			
a) Hardware (Lumber, Nails, etc.)			
b) Fabric, Canvas, Plaster			
Total Materials Costs			
6. Transportation			
a) Air Travel (itemized by trip)			
b) By Car (estimated per mile)			
Total Transportation Costs			
7. Documentation and Marketing			
a) Printing and Copying Costs			
b) Web Design and Development			
c) Social Media			
d) Marketing			
Total Documentation and Marketing Costs			

Pre-Profit Estimated Costs	
Profit or Fee	
Total Costs	

If your direct labor requires any special training or you need to hire new employees for a particular project, you should build those costs into this part of the budget. Charging a customer or funding source for training your employees might seem odd, but that money needs to come from somewhere. If a customer or client wants a special service, they should be willing to pay you to hire and train employees.

Your accountant or business managers will also need to calculate *labor overhead* for any hourly employees on the project. Labor overhead includes fringe benefits, insurance, vacations, sick leave, and retirement. Overhead is typically calculated from standard formulas that an accountant will have available. Again, it is best to leave these calculations to the accountants.

Indirect Labor

Indirect laborers are the people who support employees who assemble the products or provide the services. Indirect laborers might include salaried supervisors, repair and janitorial services, and office and clerical support. In a budget, indirect labor is usually broken down into subcategories and charged on a salaried or hourly basis. The costs of salaried labor, especially for larger projects, are estimated similarly to management costs, meaning salaries are multiplied by the percentage of time these employees will spend on the project.

Hourly indirect labor is handled similarly to hourly direct labor. To calculate hourly indirect labor, estimate how many hours your repair staff, janitors, office, and clerical workers will devote to a specific project. In the budget, you can list these subcategories and their costs separately (Figure 9.1 and Figure 9.2).

As with direct labor, overhead should be calculated by an accountant and included with the costs for indirect labor.

Facilities and Equipment

The facilities and equipment section of the budget includes the fixed and semi-variable costs of the buildings, equipment, and machinery that will be purchased, leased, or used during the project. Facilities costs also include any construction, renovation, or retooling of your factories, laboratories, or workspaces to accommodate the project's needs. You should include expenses like insurance and maintenance supplies in this part of the budget.

Any machines, tools, office or laboratory equipment, desks and chairs, and computer hardware or software leased or purchased would also be listed under this part of the budget. Identify the costs for larger pieces of equipment separately in the itemized budget and group smaller items into subcategories (Figure 9.1 and Figure 9.2). In most cases, it is assumed that you (the bidder or grantee) will keep the purchased equipment and tools after the project is completed. Nevertheless, you should clarify with the customer or funding source who will keep the equipment and tools after the project is completed.

In this section of the budget, your accountant should also include the depreciation costs for preexisting and purchased facilities and equipment used for the project. Depreciation is the lost value of facilities and equipment due to wear and aging during the project. Your company or organization is entitled to charge that depreciation to the customer or funding source. Depreciation is calculated according to a general formula your accountant will have available. As with overhead, let your accountant handle these calculations.

Direct Materials

Direct materials include the materials that will make up the products or services you are developing for the customer. For example, if you are proposing to produce a bike, the direct materials might include metal tubing for the frame, tires, tape for the handlebars, and so on. In an office environment, direct materials might include copying and other supplies that go into producing training manuals or marketing documents at a convention. If you are a lab researcher, direct materials involve any compounds or chemicals used in making the product or materials that will be disposed of when the project is complete. Essentially, direct materials are the items that will be either put in the product or service or disposed of after the project is completed.

In your budget, list the direct materials that will go into the products and services. Then, multiply the costs of these direct materials by the number of products you will produce (Figure 9.1 and Figure 9.2).

Remember to include documentation for the product in your calculations of direct material costs. If your deliverables include a support website, how-to videos, or a user manual, make sure you include those costs in the direct materials part of the budget.

Indirect Materials

Indirect materials are the items used in the process of turning directing materials into the product, but they aren't in the product itself. For example, cleaning supplies or lubricants for machinery are indirect materials because they are materials used while producing the product. For an office, supplies like paper, pens, pencils, staplers, and toner for the copy machine might be put in this indirect materials category (Figure 9.1 and Figure 9.2).

The section handling indirect materials is often the forgotten part of a budget, and companies and non-profit organizations have lost money because they didn't account for these kinds of items. Supplies like paper, staples, and toner might sound like small things, but their costs add up quickly.

Travel

Travel costs are often hard to estimate accurately, especially air travel costs. If your project involves travel for training, research, supervision, meetings, or conferences/conventions, you should build these costs into the budget.

> **Flight**—For travel by air, estimate how many trips will be needed. Then, contact a nationally recognized airline and ask about their standard business rate for travel to and

from the places you will need to travel to. Avoid using an airline's sale prices to estimate travel costs because the price may increase before you or your team members need to fly.

Train or Bus—Travel by train or bus is easier to estimate because the rates do not fluctuate as wildly as airline rates. When calculating these costs, contact the bus or train company and ask for their standard business rate for travel to and from the places you will need to travel. Then, enter those total costs into the budget.

Car—Travel by car (yours or your company's) is charged by the mile. Most organizations have a set rate for reimbursing mileage. Otherwise, the U.S. government publishes a standard rate that is a low but widely accepted figure. When estimating a project's mileage, you can use Google Maps or Apple Maps to figure out how many miles you or your colleagues might be asked to travel during the project. Keep in mind that the actual mileage usually ends up higher than the estimated mileage from Point A to Point B, so multiply that number by 5-10 percent. Then, multiply those miles by the per-mile rate provided by your organization or the government. If you plan on renting cars, you should only charge your customers or the funding source the cost of the rental and the gas. With rental cars, you should not charge mileage.

Lodging—If you need to stay at any hotels or other kinds of short-term lodgings, make sure you build those costs into the travel portion of the budget. To estimate these costs, check the website of a major hotel near the place you might stay. Look for their standard *rack rate* for a night. The rack rate is usually higher than the rate they charge their guests, but this rate can be documented for the customer, client, or funding source. If you stay at the hotel for less than the standard room rate, you can use the savings to offset other travel costs, which are often higher than you initially expected.

Don't underestimate your travel costs. These costs can fluctuate significantly, especially with the volatile costs of airfare, so don't budget for the cheapest possible rates. If you underestimate travel costs, those funds will need to come out of somewhere else in the budget. That can severely harm the project.

Communication and Marketing

A couple of other hidden project expenses are communication and marketing costs. Even though the Internet may seem free to the average employee, the Internet service provider (ISP) is a fixed cost your company or organization must pay. Your accountant should be able to estimate how much a project's Internet usage, phone usage, and postage will cost. If you don't have an accountant, you may need to look at your records from past projects to help you estimate how much it will cost to communicate with others.

Communication also includes any costs to disseminate your findings, especially with grant-funded projects. Funding sources will want any discoveries or new concepts sent out to journals and magazines or made available on websites. Some journals charge publication fees to offset their costs. If the funding source wants you to communicate with others, build those costs into this part of the budget.

Any marketing related to your project can also be listed in this area. You may need to pay someone to develop webpages or an entire website for your project. You may also need someone to develop a social media campaign or manage the social media interactions that may be part of the project. Do not try to lighten the budget by attempting to manage social media marketing and website development yourself. Website and social media specialists are professionals and know how to do these things!

Projects with more comprehensive marketing and advertising campaigns will need their own budget. If your project requires comprehensive marketing and advertising, contact your company's or organization's accountant for guidance and discuss the public relations campaign with your marketing or public relations teams. They can help you estimate the costs.

Profit (For-Profit Business Proposals Only)

Your company deserves to make a profit for providing your products or services to your customers or clients. In proposals, profits are handled in two ways: First, the profit can be built into the costs of the products or services. For example, let's say one unit of a company's product costs $50 to produce. The budget submitted to the customer or client charges $55 per unit. In this case, the bidder has chosen to use the product as the source of the profit because each one will yield $5 in profit.

The second way to handle profit, especially when providing the customer with an itemized budget, is to include a profit line in the budget (Figure 9.1). If you submit a proposal to a government agency, you will often be required to disclose your profit. The profit becomes an additional charge based on a flat rate or a percentage of the actual costs. An accountant should be able to help you develop a fair estimate of the profit you should request in your budget.

There are advantages to both ways of calculating profit. Building the profit into the product or service costs is a hidden way to create a profit margin in a proposal. That said, you might find yourself defending the price of your $70 per-unit product when the customer or client argues it should only cost $60 to produce.

The advantage of a separate profit line is that it clearly shows the customer or clients how much money beyond the costs you would like to make for your products or services. However, the profit line is also a prime target for negotiation. Your customer or client may see that you have listed a 15 percent profit for the project and try to negotiate a lower percentage.

In most corporate proposals, profit is not expressed directly in the budget. Companies often consider their profit calculations confidential and are reluctant to submit them to the customer. Also, some larger companies consider their profit margin nonnegotiable, even though everything in a proposal is ultimately open for negotiation.

Figure 9.2: A Budget Worksheet for a Non-Profit Project

Itemized Budget (Non-Profit Organization)

	Estimated Hours or Miles	Rate per Hour or Mile (if applicable)	Other Funds or In-Kind Contributions	Estimated Total Cost
1. Management				
a) Project Manager or Principle Investigator				
Benefits				
b) Managers and Supervisors (list out)				
Benefits				
c) Specialists				
Benefits				
d) Consultants				
e) Additional Training				
Total Management Costs				
2. Labor				
a) Direct labor				
b) Indirect labor				
c) Contractors				
d) Training				
e) Labor Overhead				
Fringe Benefits				
Insurance				
Vacation				
Sick Leave				
Pension				
Total Labor Costs				
3. Equipment and Supplies				
a) Computer Hardware and Software				
b) Tools				
c) Tables, Chairs, Desks				
d) Office Supplies				
e) Depreciation of Equipment				
Total Equipment Costs				

Figure 9.2: A Budget Worksheet for a Non-Profit Project (Cont.)

	Estimated Hours or Miles	Rate per Hour or Mile (if applicable)	Other Funds or In-Kind Contributions	Estimated Total Cost
4. Facilities and Maintenance				
a) Building Rent				
b) Retooling Facilities				
c) Depreciation of Facilities				
d) Insurance				
e) Internet and Telephone				
f) Water, Electricity, and Gas				
g) Maintenance				
Total Facilities Costs				
5. Materials				
a) Direct Materials (Lumber, Nails, etc.)				
b) Indirect Materials (Lubricants, etc.)				
Total Materials Costs				
6. Transportation				
a) Air Travel (itemized by trip)				
b) By Car (estimated per mile)				
Total Transportation Costs				
7. Documentation and Marketing				
a) Printing and Copying Costs				
b) Web Design and Development				
c) Social Media				
d) Marketing				
Total Documentation and Marketing Costs				

Pre-F&A Estimated Costs	
Multiplied by F&A Rate	
Total Costs	

Again, there is no substitute for a good accountant when working out these issues in a budget. An accountant can help you choose the best way to express the profit in your budget. And remember, you will need to commit to the figures detailed in your proposal (maybe even in court), so it is in your best interest to let an expert help you with the budget.

Facilities and Administrative Costs (F&A) (Grant Proposals Only)

Non-profit organizations, especially universities, will often include a budget line called facilities and administrative costs (F&A) or *indirect costs* (Figure 9.2). These calculations cover the overhead costs of your university or non-profit organization. At many research universities, these costs are about half the pre-F&A costs for any project completed on campus. In other words, after all the project's costs are calculated, they are multiplied by about 50 percent. That additional amount is then added to the budget. F&A or indirect costs for off-campus projects are usually calculated at a lower rate.

These costs are generally not under your control, so let your organization's accountant figure them out. However, you should keep these costs in mind because they will significantly inflate the total costs of your project.

Cost Sharing (Grant Proposals Only)

Another wrinkle in a grant proposal budget is the opportunity (or requirement) for cost sharing, meaning your organization or another funding source will share some of the project's costs. These costs are usually handled with matching funds and in-kind contributions (Figure 9.2).

Matching Funds

A funding source may require your organization to fully or partially match the funds requested from them. For example, a foundation may ask you to match their grant one-to-one, meaning you need to match each of the foundation's dollars with a dollar from your organization or another source. Or a funding source may tell you they will only pay 50 percent of equipment costs. If so, your organization would be responsible for supplying or finding the other 50 percent to pay for those costs.

Non-profit organizations raise matching funds in a few ways:

- Withdraw cash from an endowment or similar fund
- Promise to do a fundraiser if the grant is awarded
- Match the funds with grants from other funding sources

Matching with cash can be problematic for non-profit organizations, so you should be careful about any grant opportunity that requires a cash match. A funding source may award you a large grant, but you need to be realistic about whether your organization can raise the cash to match that amount.

In-Kind Contributions

In-kind contributions are non-cash gifts and services that can often be used to match the funds provided by a funding source. These contributions can come in many forms, including volunteer time, donated goods and services provided by supporting businesses, and donated space. For example, you can calculate the value of your volunteers' time (usually by the hour) and use that time as cost sharing. Similarly, if supporting businesses donate goods and services, such as materials or accounting services, you can add those to your in-kind contribution.

Some excellent sources for in-kind contributions are your facilities themselves. If your organization has a facility that is already paid for, you can often use that facility's costs or potential rent as an in-kind contribution. For example, let's say a gallery in your building will house a display. You can calculate that gallery's potential rent and upkeep to use as an in-kind contribution. No money is actually changing hands because your organization already owns the building, and you already pay for the electricity, heat, water, and janitorial services. Nevertheless, you can often use those costs toward the match. As always, an accountant can help you figure this out.

In your budget sheet, cost sharing can be represented in two ways. You might include a line in the budget called "cost sharing" in which you sum up all the matching and in-kind contributions. Many non-profits also like to add a column in their budget sheet that helps them itemize how any costs will be shared.

Even if the funding source does not explicitly ask for cost sharing, it's a good practice to show that your organization will find ways to share the cost burden. Cost sharing shows your organization's financial commitment to the project, which will make funding sources more comfortable about putting up their funding.

Writing the Budget Rationale or Budget Section

The budget will usually appear as a stand-alone section inside the proposal or in an appendix. Some proposal writers prefer to sandwich a Budget section between the Qualifications section and the proposal's conclusion. Other proposal writers prefer to include the budget and a budget rationale in an appendix.

Positioning the Budget

There are, of course, pros and cons to both approaches. A Budget section with a budget rationale placed inside the body of the proposal will urge the readers to look over your cost figures. However, a Budget section may run a few or multiple pages, bogging down the readers in numbers and tables right when you want them to think about the benefits of your project.

Placing the budget with a budget rationale in an appendix is a good way to avoid slowing down and distracting the readers. However, readers will sometimes complain that placing a budget in the appendix is designed to "hide" the costs, even if you provide a summary of the costs in the conclusion.

You might ask the readers where they would like the budget placed. If they don't have a preference, we suggest putting it in the appendix, so an extended review of the budget doesn't kill the momentum of your proposal.

Organizing the Budget Rationale

Wherever you choose to include the budget, you should think of it as an argument that stands on its own. It is usually not sufficient to say, "Here is our budget," as though a table with an itemized list of expenses speaks for itself. Instead, you should explain the reasoning behind your figures.

Like any section in a proposal, the Budget section or rationale has an opening, body, and closing. The opening paragraph identifies the topic of the section (the budget) and the purpose of the section (to present and discuss the project's costs). It should also include a main claim the Budget section will support or prove. For example, you might claim, "By keeping expenses low, we offer the most efficient fabrication services in this industry." This kind of claim will focus your discussion of the budget, giving the section something to prove.

The body of the Budget section backs up that kind of claim with a table of expenses and explanatory remarks that highlight the important parts of the budget. As much as possible, your budget rationale should try to anticipate some of the readers' questions about the costs. Most readers are not as familiar with a project's equipment, facilities, and materials as the company or organization submitting the proposal, so some additional explanation can go a long way toward helping the readers feel confident about the projected costs of the project.

A table of budget figures, like the ones shown in Figure 9.1 and Figure 9.2, is the heart of a Budget section or rationale. Your budget rationale should refer to specific lines in the budget table, explaining how and why the figures were calculated in a particular way.

The closing of a Budget section or rationale should restate your main point in the opening paragraph. The closing, usually only a sentence or two, is designed to round off the section and reinforce the soundness of your cost estimates. You can also mention that you or someone from your company or organization is available to walk the readers through the figures if they have questions.

Projecting Confidence

When writing the budget rationale, you should suppress any urges to be apologetic or defensive about the costs of your proposal. The budget is where you need to sound confident. If you sound uncertain or apologetic, your readers will begin to doubt the soundness of your figures. They will question your company's or organization's abilities to complete the project without exceeding the budget. Your best approach is to just discuss the budget in a straightforward and confident way. Remember: the budget needs to be accurate and acceptable to all parties. Your proposal may temporarily serve as a contract until a formal agreement is signed, so it needs to be able to stand up to close scrutiny from customers, clients, or funding sources.

Chapter Summary and Looking Ahead

Never underestimate the importance of the budget in a proposal or grant. Imagine that you are deciding whether you will buy something online. You closely evaluate the product's merits and consider its drawbacks. You look around to see if you're getting the best or at least a reasonable price. The last question you answer is, "Is this worth the asking price?" The more information you have about the product or service, the more likely you will say yes.

Readers evaluate proposals and grants in much the same way. They will review the whole proposal, but their decision will be driven by the money. The more information the readers have about how the costs were determined, the more likely they are to make a firm decision about whether your proposed plan suits their needs.

The following chapters will discuss the style and design of proposals and grants. If you've been drafting your proposal or grant by following the chapters in this book, you are close to finished. It's time to revise and design the proposal to make it easier to read, more attractive, and more persuasive.

Try This Out!

1. Using your personal finances, divide your expenses into fixed, variable, and semi-variable costs. List these costs in a budget table in which you show yourself as management or labor (depending on whether you receive a salary or are paid hourly). Then, write a one-page budget rationale to a member of your family or a friend in which you justify the major expenses in your budget.

2. Find a proposal or grant that includes an itemized budget on the Internet, at your workplace, or on your campus. Then, answer the following questions:
    a. Where does the budget appear in the proposal?
    b. How does the budget placement shape the reading of the whole proposal?
    c. How does the proposal's budget itemize costs?
    d. Does the proposal include a rationale that explains the features of the budget to the readers?
    e. If not, do you think a budget rationale would be helpful?

3. Write a budget for a practice or actual proposal or grant. First, decide whether you need a fixed or flexible budget. Second, list your major costs, dividing them into fixed, variable, and semi-variable categories. Third, place these costs in a Budget Worksheet, breaking them down according to the larger categories described in this chapter. Finally, write a short budget rationale in which you lead the readers through the highlights of your budget.

4. Ask your local Small Business Association (SBA) or Chamber of Commerce (CoC) representative about the budget requirements for a business plan to start a new business. Ask

the SBA/CoC representative what kinds of information investors or banks need to consider before lending money.

5. Ask a grants officer on your college campus about the process for writing a budget for a grant. Ask how their office prefers grant writers to be involved in the budgeting process. Ask about the typical mistakes in budgeting made by grant writers. When you are finished, report your findings to your class.

Case Study: The Carbon Neutral Campus Project—How Much Will This Cost?

George volunteered to take on the difficult task of working out a budget for the Carbon Neutral Campus grant proposal. He had learned long ago that bringing in a financial specialist as early as possible was a good idea, so he scheduled a meeting with Bill Vonn, the development officer for the College of Engineering at Durango University. In the past, George had written successful grant applications to the Department of Energy (DOE) and the National Science Foundation (NSF), but he had never sent a grant proposal to a private foundation. So, he was uncertain about how to handle the budgetary issues.

When George called him, Bill was enthusiastic about the Carbon Neutral Campus Project. Durango University hadn't sent a grant application to the Tempest Foundation for six years, so he was eager to re-establish connections with this funding source. He thought the grant had a good chance of being funded.

When they met, George said, "I'm unsure how we divide these costs. What will the foundation pay for, and what costs will they reject?"

Bill smiled, "George, I think you're getting ahead of yourself. Let's go through the grant and identify the specific costs of the project."

George and Bill began underlining the major costs of the project:

- Fee for an urban planner or urban planning firm
- Salaries for the project leaders and participants in the project
- Rental of ballroom and breakout rooms for the charrette
- Promotional materials and activities to invite stakeholders to the charrette
- Supplies for the charrette
- Food and refreshments for the charrette
- A room at the library to archive and store materials
- A webmaster to develop and maintain a website
- Projectors for presentations at the charrette and other meetings
- Equipment for creating podcasts of the meetings
- Postage to send items to the Tempest Foundation
- Costs for the Internet provider
- Travel costs for project leaders, urban planners, and project reviewers

- Costs to disseminate results of the project
- Honoraria for project evaluators

George looked at the list. He shook his head and said, "There's no way the Tempest Foundation is going to pay for all these things."

Bill agreed, "You're right. But that's not all bad. The university will need to pay for some of these items, but we can put most of those costs in the cost-sharing and in-kind columns in the budget. We can also count items like salaries and volunteer time toward cost sharing."

"I see," said George. "Writing a grant to a foundation is slightly different than writing a grant to the DOE or the NSF."

"True. Private foundations will expect the university to foot some of the bill for this project. We can do much of the cost-sharing by taking advantage of stuff we're already paying for, like the rent for the ballroom in the union."

"Rent?" asked George.

"Yeah, the ballroom would cost $1000 per day to rent. So, we can multiply that figure by five days and put that figure as an in-kind cost on our side of the ledger."

"That's helpful."

Bill added, "And if President Wilson is behind this project, he'll allocate funds from the university's endowment into the budget. That will show the Tempest Foundation that the university is serious about this project."

Bill pulled out a Budget Worksheet and began filling it out. After crunching the numbers, Bill and George wrote a tentative budget (Figure 9.3) and budget rationale.

As Bill added up the numbers, George began getting nervous again. "Wow, that added up quick," he said.

"It always does," said Bill. "Private foundations don't have bottomless pockets of money, but they are willing to dig deeper for high-impact projects. If they like this project, the Tempest Foundation will be willing to invest a significant amount of money. If they can't afford to give us the full amount, they will tell us what they can afford. At that point, we may need to find donors who can match. That will make up the difference."

"What about F&A costs? Will they pay those?" asked George.

"I doubt it. The university will need to waive those costs or try to use them as cost-sharing. This project isn't resource intensive, so persuading the university to waive F&A shouldn't be too difficult."

"We'll keep our fingers crossed."

Bill put a copy of the budget sheet and rationale in Google Drive. He said, "George, finish writing up your budget, and I'll set up a video conference with the university's head development officer, Janet Freund. She can tell us how much the university can put into this project and how much additional donor money we need to raise. Then I will call the point of contact at the Tempest Foundation to see how they feel about some of these costs."

"Thank you. I really appreciate your help," said George as Bill stood up and started walking toward the door.

"Hey, this is my job." Bill laughed, "I'm just glad you didn't call me a few days before the grant's due. The extra time gives me some flexibility to work with the university and the foundation. Your project looks great. Let's keep our fingers crossed the Tempest Foundation likes this idea."

Figure 9.3: Bill's Rough Budget for the Carbon Neutral Campus Grant

Item	Tempest Foundation	Durango University
Project Leaders' Salaries and Benefits		
George Tillman (20 percent time)		$41,300.00
Anne Hinton (10 percent time)		28,240.00
Major Participants' Salaries and Benefits		
Diane Smith, Urban Planner	$24,000.00	
Gina Sanders, Librarian (10 percent time)		13,600.00
Staff and Assistants		
3 facilitators	30,000.00	
5 research assistants (50 percent time)	22,900.00	
Clerical assistance (100 hours @ 15.00/h)	18,000.00	6,000.00
Volunteers (200 hours @ 15.00/h)		3,000.00
Services and Facilities		
Rental of a ballroom in the student union		5,000.00
Room at the library to house materials		1080.00
Projector rental		900.00
Podcasting equipment rental		400.00
Materials		
Supplies for charrette	860.00	
Promotional Materials	1960.00	
Food and Refreshments		8,000.00
Travel and Housing Expenses		
Project leaders	6,400.00	
Urban planners	6,000.00	
Evaluators	4,000.00	
Communications and Dissemination		
Internet Service Provider	460.00	
Postage and Shipping	620.00	
Documentation	1860.00	
Evaluation		
Honoraria for 2 evaluators	2,000.00	

Item	Tempest Foundation	Durango University
Indirect Costs (F&A)		
Total Costs	$109,060.00	$107,520.00

If you would like to see how the Budget of the Carbon Neutral Campus proposal turned out, please go to Appendix A.

10 WRITING WITH STYLE

Choosing to Write Clearly and Persuasively

Style is a critical feature of any proposal. Your style expresses the attitude of your company or organization toward the project. Style also reflects your character by signifying the relationship your company or organization wants to build with the readers. In a word, style is about quality. It showcases your commitment to excellence and the attention to detail in your company or organization. Style operates on a few different levels in a proposal. On the sentence level, good style involves choosing the right words or forming sentences that are easy to read. On the paragraph level, style involves weaving sentences together in ways that emphasize your main points and lead the readers comfortably through your ideas. At the document level, style involves setting an appropriate tone and weaving themes into your work that appeal to your readers' emotions and values.

Your stylistic choices do more than make the content easier to read and more persuasive. Your choice to write clearly and persuasively illustrates your clear-headedness, your emphasis on quality, and your willingness to communicate and work with the readers. Style helps you achieve credibility, or *ethos*, with your readers.

Good style is a choice you can and should make. This chapter will discuss two types of style that are prevalent in proposals and grants: Plain style and persuasive style.

- Plain style uses straightforward sentences and paragraphs that express your ideas clearly to the readers.
- Persuasive style motivates your readers by appealing to their emotions and values.

Both the plain and persuasive styles have their place in any given proposal. The challenge is to balance these two styles in ways that will inform the readers and move them to say yes to your proposal or grant.

Artificial intelligence (AI) applications are especially helpful with editing sentences and paragraphs for plain style. They are especially good at proofreading for grammar and punctuation. The discussion of writing plain sentences that follows will show you the techniques editors and AI applications use to make sentences easier to read and clearer to readers.

But you should use AI cautiously. These applications can create a "professional" or "informal tone" with surface changes to words and phrases, but using more advanced features of persuasive style, such as metaphors or personification, is much more difficult for these applications. Readers often complain about the artificial tone of AI-generated content, which can sound empty, overly formal, or too enthusiastic. Instead, by using plain and persuasive style techniques, you can create an authentic tone that makes a positive connection with your readers.

Writing Plain Sentences

As a high school student, you were more than likely advised to "write clearly" or "write in concrete language," as though simply making up your mind to write clearly or concretely was all it took. Truth is, writing plainly is a skill that requires practice and concentration. Fortunately, once you have learned a few simple copyediting guidelines, writing plainly will become a natural strength in your writing.

The techniques you will learn in this section are also used by AI applications to make writing clearer and easier to read. AI can do some wordsmithing for you, but knowing what the AI application is doing and why is helpful. That way, you can write clearer sentences to begin with. Also, if the AI application edits your writing in a way that sounds wrong or confusing, you can step in to copyedit those sentences yourself.

To start, let's consider the parts of a basic sentence. As far back as elementary school, you learned that a sentence typically has three main parts: a subject, a verb, and a comment. The *subject* is what the sentence is about. The *verb* is what the subject is doing. And the *comment* says something about the subject. For example, consider these three variations of the same sentence.

Subject	Verb	Comment
The Institute	provided	the government with accurate crime statistics.
Subject	Verb	Comment
The government	is provided	with accurate crime statistics by the Institute.
Subject	Verb	Comment
Crime statistics	were provided	to the government by the Institute.

Notice that the content in these three sentences has not changed, but the emphasis in each sentence changes as different nouns are put into the subject. Sentence A is about the "Institute." Sentence B is about the "government." Sentence C is about "crime statistics." By changing the subject of the sentence, we essentially shift the focus of the sentence, drawing our readers' attention to specific issues.

This simple understanding of the different parts of a sentence is the basis for eight guidelines you can use to write plainer sentences in proposals and grants (Figure 10.1).

Figure 10.1. Eight Guidelines for Plain Style

> **Guideline 1**: The subject should be what the sentence is about.
> **Guideline 2**: Make the "doer" the subject.
> **Guideline 3**: State the action in the verb.
> **Guideline 4**: Put the subject early in the sentence.
> **Guideline 5**: Eliminate nominalizations.
> **Guideline 6**: Avoid excessive prepositional phrases.
> **Guideline 7**: Eliminate redundancy.
> **Guideline 8**: Make sentences "breathing length."

Guideline 1: The Subject Should Be What the Sentence Is About

At a basic level, weak style usually happens when the readers cannot easily identify what the sentence is about. For example, what is the subject of the following sentence?

1. Ten months after the Hartford Project began, in which a team of our experts conducted close observations of management actions, it is our conclusion in the end that the scarcity of monetary funds is at the basis of the inability of Hartford Industries to appropriate resources to essential projects that have the necessities that are greatest.

This sentence is difficult to read for several reasons, but the most significant problem is the lack of a clear subject. What is this sentence about? The word *it* is currently in the subject position, which is why readers will have trouble figuring out what the sentence is about. The sentence could be about the *conclusion, experts, the Hartford Project*, or *scarcity of monetary funds*. Many other nouns also seem to be competing to be the subject of the sentence, such as *observations, management, structure, inability,* and *company*. These nouns bombard the readers with potential subjects, undermining their efforts to figure out what the sentence is about.

When the sentence is restructured around *experts* or *scarcity*, most readers will find it easier to understand:

1a. Ten months after the Hartford Project began, *our experts* have concluded through close observations of management actions that the scarcity of monetary funds is at the basis of the inability of Hartford Industries to appropriate resources to essential projects that have the greatest necessity.

1b. According to our experts' observations of management actions ten months after the Hartford Project began, *the scarcity of monetary funds* is at the basis of the inability of Hartford Industries to appropriate resources to essential projects that have the greatest necessity.

Both sentences are still difficult to read. Nevertheless, these versions are easier to read than the original because the noun occupying each sentence's subject slot is the focus of the sentence (i.e., what the sentence is about). Sentence 1a is about "our experts," and sentence 1b is about "the scarcity of funds." We will return to this sentence about Hartford Industries after discussing the other guidelines for plain style.

Guideline 2: Make the "Doer" the Subject

Which revision of sentence 1 (1a or 1b) is easier to read? Most people would point to sentence 1a, where the *experts* are in the subject slot. Why? In sentence 1a, the experts are doing something. In sentence 1b, *scarcity* is an inactive noun that isn't doing anything. Whereas experts take action, *scarcity* is just something that happens.

Readers tend to focus on who or what is doing something in a sentence. To illustrate, which of these sentences is easier to read?

2a. On Saturday morning, the paperwork was completed in a timely fashion by Jim.

2b. On Saturday morning, Jim completed the paperwork in a timely fashion.

Most people would say sentence 2b is easier to read. *Jim*, the subject of the sentence, is doing something, while the *paperwork* in sentence 2a is inactive. It's not doing anything.

The active person or thing usually makes the best subject of the sentence.

Guideline 3: State the Action in the Verb

Similarly, Guideline 3 states that the verb should contain the action in the sentence. Once you have determined who or what is doing something, ask yourself what that person or thing is doing. Find the action in the sentence and make it the verb. For example, consider these sentences:

3a. The detective investigated the loss of the payroll.

3b. The detective conducted an investigation into the loss of the payroll.

3c. The detective is the person who is conducting an investigation into the loss of the payroll.

Sentence 3a is easier to understand because the sentence's action is expressed in the verb. Sentences 3b and 3c are increasingly more difficult to understand because the action, *investigate*, is further removed from the verb slot of the sentence. A sentence is clearer when the verb (*investigate*) states the action in the sentence.

Guideline 4: Put the Subject Early in the Sentence

Subconsciously, readers start every sentence looking for the subject. The subject anchors the sentence because it tells the readers what the sentence is about. So, if the subject is buried somewhere in the middle of the sentence, your readers will have greater difficulty finding it, and the sentence will be harder to read. Consider these two sentences:

4a. If deciduous and evergreen trees experience yet another year of drought like the one observed in 1997, the entire Sandia Mountain ecosystem will be heavily damaged.

4b. The entire Sandia Mountain ecosystem will be heavily damaged if deciduous and evergreen trees experience yet another year of drought like the one observed in 1997.

The problem with sentence 4a is that it forces the readers to hold all those details (i.e., trees, drought, 1997) in short-term memory before the sentence finally identifies its subject. Readers almost feel a sense of relief when they find the subject because until they locate it, they cannot figure out what the sentence is about.

Quite differently, sentence 4b tells the readers what the sentence is about up front. With the subject placed early in the sentence, the readers immediately know how to connect the comment with the subject.

Long introductory or transitional phrases tend to make sentences harder to read. Your sentences will be easier to read if you avoid these long phrases at the beginning of sentences.

Guideline 5: Eliminate Nominalizations

Nominalizations are perfectly good verbs and adjectives that have been turned into awkward nouns. They frequently end with *–ion*. For example, look at these sentences:

5a. Management has an expectation that the project will meet the deadline.

5b. Management expects the project to meet the deadline.

In sentence 5a, *expectation* is a nominalization. Here, a perfectly good verb, *expect*, is being used as a noun, *expectation*. After turning the nominalization into a verb, sentence 5b is much shorter than sentence 5a. It also has more energy because the verb, *expect*, is now an action verb.

Consider these two sentences:

6a. Our discussion about the matter allowed us to make a decision on the acquisition of the new X-ray machine.

6b. We discussed the matter and decided to buy the new X-ray machine.

Sentence 6a contains three nominalizations: *discussion*, *decision*, and *acquisition*. These nominalizations make the sentence hard to understand. The revised version, sentence 6b, turns all three nominalizations into simpler verbs, making the sentence much easier to understand.

The energy added to the sentence is another benefit of changing nominalizations into verbs. Nouns feel inert to the readers, while verbs add action and energy.

Why do writers use nominalizations in the first place? First, humans generally think in nouns, so our first drafts are often filled with nominalizations. While revising, an effective writer will turn those first-draft nominalizations into action verbs. Second, some people believe that using nominalizations makes their writing sound more formal or important. In reality, nominalizations usually make sentences harder to read, not more formal or important.

Guideline 6: Avoid Excessive Prepositional Phrases

Prepositional phrases are necessary in writing, but they are often overused in ways that make sentences feel long and tedious. Prepositional phrases start with prepositions (e.g., *in, of, by, about, over, under*), and they are used to modify nouns. For example, in the sentence "Our house by the lake in Minnesota is lovely," the phrases *by the lake* and *in Minnesota* are prepositional phrases. They modify the nouns *house* and *lake*.

Prepositional phrases are fine when used in moderation but become problematic when used in excess. For example, in sentence 7a the prepositions have been italicized and prepositional phrases underlined. Sentence 7b a revision of 7a with fewer prepositional phrases:

7a. The decline *in* the number *of* businesses owned *by* locals *in* the town *of* Artesia is a reflection *of* the increasing hardships faced *in* rural communities *in* the Southwest.

7b. Artesia's declining number of locally owned businesses reflects the increased hardships faced by Southwestern rural communities.

You don't need to eliminate all the prepositional phrases in a sentence. Instead, look for places where prepositional phrases are chained together. Then, try condensing the sentence by converting some prepositional phrases into adjectives. For example, in sentence 7b the phrase *in the town of Artesia* was reduced to one word, *Artesia's*. The prepositional phrases *in rural communities in the Southwest* were reduced to *southwestern rural communities*. The revised version of the sentence, 7b, is much shorter and easier to read.

Guideline 7: Eliminate Redundancy

When trying to stress an important point in a proposal, writers sometimes use redundant phrasing. For example, they might write *unruly mob*, as though some mobs are orderly, or they might talk about *active participants*, as though someone can participate without doing anything.

Sometimes buzzwords and jargon lead to redundancies like, "We should collaborate together as a team" or "Empirical observations will provide a new understanding of the subject." In some cases, we might use a synonym to modify a synonym by saying something like, "We suggested important, significant changes." Redundancies should be eliminated because they use two or more words to do the work of one. As a result, the readers need to work twice as hard to understand one basic idea.

Guideline 8: Make Sentences Breathing Length

A natural sentence can be spoken in one breath. The period at the end of each sentence signals the readers to breathe. Of course, when reading silently, people do not actually breathe when they come across a period. Nevertheless, readers do take a mental pause at the end of each sentence as though taking a breath. A sentence that runs on and on forces readers to mentally hold their breaths. By the end of an especially long sentence, we are more concerned about getting through it than deciphering it.

The best way to think about sentence length is to imagine how long it takes to comfortably say a sentence out loud. If the written sentence is too long to say out loud in one breath, it needs to be shortened or cut into two sentences. After all, you don't want to asphyxiate your readers.

On the other hand, if the sentence is very short, maybe it needs to be combined with one of its neighbors to make it a more comfortable breathing length. You also want to avoid hyperventilating the readers with a string of short sentences.

A Simple Method for Writing Plainer Sentences

As a recap, here is a process for writing plainer sentences. First, write your draft as usual, paying little attention to the style. Then, as you revise, read the draft aloud to identify difficult sentences and apply the six steps shown in Figure 10.2.

Figure 10.2: Six Steps to Achieve Plainer Writing

> Step 1. Identify who or what the sentence is about.
> Step 2. Turn that who or what into the subject and move the subject to an early place in the sentence.
> Step 3. Identify what the subject is doing and move that action into the verb slot.
> Step 4. Eliminate prepositional phrases, where appropriate, by turning them into adjectives.
> Step 5. Eliminate unnecessary nominalizations and redundancies.
> Step 6. Shorten, lengthen, combine, or divide sentences to make them breathing length.

With these six steps in mind, let's revisit sentence 1, the example of weak style offered at the beginning of our discussion of plain style:

Original

1. Ten months after the Hartford Project began, in which a team of our experts conducted close observations of management actions, it is our conclusion in the end that the scarcity of monetary funds is at the basis of the inability of Hartford Industries to appropriate resources to essential projects that have the necessities that are greatest.

Revision

1a. After a ten-month study, our experts concluded that Hartford Industries' budget shortfalls have limited its support for priority projects.

In this revised version, the subject *(our experts)* was moved into the subject slot, and it was moved to an early place in the sentence. Then, the sentence's action *(concluded)* was moved into the verb slot. Prepositional phrases like *to appropriate resources to essential projects* were turned into adjectives. Nominalizations like *conclusion* and *necessity* were turned into verbs or adjectives. And finally, the sentence was shortened to breathing length. The resulting sentence still offers the same content to the readers—just more plainly.

Writing Plain Paragraphs

Writing plain paragraphs is also a skill that you can learn. Some simple methods are available to help you write paragraphs that will help your readers find and remember the important points in your proposals and grants.

The Elements of a Paragraph

You may have been taught that paragraphs are primarily about length: "Once you get to four or five sentences, you need to start a new paragraph." But paragraphs are about ideas, not length. Each paragraph should have a single controlling idea that it supports or proves.

To help readers follow the ideas expressed in paragraphs, you can use up to four kinds of sentences: a transition sentence, a topic sentence, a support sentence, and a point sentence (Figure 10.3). Each of these sentences plays a different role in the paragraph.

Transition Sentence

The purpose of a *transition sentence* is to make a smooth bridge from the previous paragraph. For example, a transitional sentence might state, "With these facts in mind, let's consider the current opportunity available." The previous paragraph provided those facts. So, by referring to the previous paragraph, the transition sentence provides a smooth bridge into the new paragraph.

Most paragraphs, however, do not need a transition sentence to make a bridge. Transition sentences are typically used when the new paragraph handles a significantly different topic than the previous paragraph. The transition sentence redirects the discussion from the previous topic to the new one.

Topic Sentence

The *topic sentence* is the claim or statement the rest of the paragraph tries to prove or support. The topic sentence expresses the central, controlling idea of the paragraph. In a proposal, topic sentences typically appear in each paragraph's first or second sentence. They are placed up front in each paragraph for two reasons. First, the topic sentence sets a goal for the paragraph to reach by telling the readers the claim you are trying to prove. Then, the remainder of the paragraph proves that claim with facts, examples, and reasoning. If the topic sentence appears at the end of the paragraph, the readers are forced to rethink all the details now that they know what the paragraph was trying to prove. For most readers, all that mental backtracking is a bit annoying.

The second reason for putting the topic sentence up front is that it is the most important sentence in any given paragraph. Since readers tend to pay the most attention to the beginning of a paragraph, placing the topic sentence up front guarantees they will read it closely. Likewise, scanning readers tend to concentrate on the beginning of each paragraph. If the topic sentence is buried in the middle or at the end of the paragraph, they will miss it.

Figure 10.3 Four Types of Sentences in a Paragraph

Transition Sentence—makes a transition from the previous paragraph

Topic Sentence—makes a statement or claim that the paragraph will support or prove

Support Sentences
- Reasoning
- Examples
- Data
- Definitions
- Descriptions
- Anecdotes

Point Sentence—states the point of the sentence

Support Sentences

The *support sentences* in a paragraph can come in many forms. Generally, there are two ways to argue logically (i.e., reasoning and examples). Sentences that use reasoning tend to make if/ then, cause/ effect, better/worse, greater/ lesser kinds of arguments for the readers. Sentences that use examples illustrate points for the readers by showing them situations or examples that support your claim in the topic sentence.

For the most part, sentences containing reasoning and examples will make up most of a paragraph's support sentences. Other support will provide facts, data, definitions, and descriptions. Support sentences are intended to support or prove the claim made in the paragraph's topic sentence.

Point Sentences

A point sentence restates the topic sentence at the end of the paragraph. Point sentences reinforce the paragraph's topic sentence by restating its original claim in new words. They are especially useful in longer paragraphs where the readers may not fully remember the claim stated at the beginning of the paragraph. These sentences often start with transitional devices like *therefore, consequently*, or *in sum* to signal to the readers that the point of the paragraph is being restated.

Point sentences are optional in paragraphs and should primarily be used when a paragraph's claim needs to be reinforced for the readers. But, use point sentences sparingly. Too many point sentences will cause your proposal to sound repetitious and even condescending to the readers.

Building a Plain Paragraph

Only the topic sentence and the support sentences are needed to build a solid paragraph. Transitional and point sentences are optional and best used when bridges need to be made between paragraphs or specific points need to be reinforced.

Here are the four kinds of sentences used in a paragraph:

8a. How can we accomplish these five goals? (**transition**) Universities need to study their core mission to determine whether distance education is a viable alternative to the traditional classroom (**topic sentence**). If universities can maintain their current standards while moving their courses online, distance education may provide a new medium through which nontraditional students can take classes and perhaps earn a degree (**support**). Utah State, for example, reports that students enrolled in their online courses have met or exceeded the expectations of their professors (**support**). If standards cannot be maintained, however, we may find ourselves returning to the traditional on-campus model of education (**support**). In the end, the ability to meet a university's core mission is the litmus test to measure whether distance education will work (**point sentence**).

8b. Universities need to study their core mission to determine whether distance education is a viable alternative to the traditional classroom (**topic sentence**). If universities can maintain their current standards while moving their courses online, distance education may provide a new medium through which nontraditional students can take classes and perhaps earn a degree (**support**). Utah State, for example, reports that students enrolled in their online courses have met or exceeded the expectations of their professors (**support**). If standards cannot be maintained, however, we may find ourselves returning to the traditional on campus model of education (**support**).

As you can see in paragraph 8b, a paragraph works fine without transition sentences and point sentences. Nevertheless, a transition sentence or point sentence can make a paragraph more readable while amplifying its point.

Aligning Sentence Subjects in a Paragraph

Have you ever needed to stop reading a paragraph because each sentence seems to go off in a new direction? Have you ever run into a paragraph that feels bumpy as you read it? More than likely, the problem with the paragraph was due to a lack of *alignment* among the paragraph's sentences. To illustrate, consider this paragraph:

9. The lack of <u>technical knowledge</u> about the electronic components in automobiles sometimes causes car owners to doubt the honesty of car mechanics. Although <u>they</u> tend to be knowledgeable about the mechanical workings of their automobiles, <u>the nature and scope of the electronic repairs</u> needed in modern automobiles are rarely understood by car owners. For instance, the <u>function and importance</u> of a transmission in a car are generally well known to all car owners, but the <u>wire harnesses and printed circuit boards</u> that regulate the fuel consumption and performance of their car are rarely familiar. <u>Repairs for these electronic components</u> can often run over $400. <u>That</u> is a large amount for a customer who cannot even visualize what a wire harness or printed circuit board looks like. In contrast, a <u>$600 charge</u> to repair the transmission on the family car is more readily understood and accepted.

There's nothing grammatically wrong with this paragraph—it's just hard to read. Why? The subjects of the sentence change with each new sentence. Look closely at the underlined subjects of the sentences in the above paragraph. The subjects are all different, causing each sentence to feel like it is striking off in a new direction. As a result, each new sentence forces the readers to shift focus to concentrate on something new.

To avoid this bumpy, unfocused feeling, you can line up the subjects so each sentence in the paragraph focuses on the same topic. To line up subjects, first ask yourself what the paragraph is about. Then, revise most of the sentences to align with that subject. Here is a revision of paragraph 9 that focuses on the "car owners" as subjects:

9a. Due to their lack of knowledge about electronics, <u>some car owners</u> doubt the honesty of car mechanics when repairs involve electronic components. <u>Most of our customers</u> are knowledgeable about the mechanical features of their automobiles, but <u>they</u> rarely understand the nature and scope of the electronic repairs needed in modern automobiles. For example, <u>most people</u> recognize the function and importance of the transmission in an automobile, but <u>the average person</u> knows very little about the wire harnesses and printed circuit boards that regulate the fuel consumption and performance of their car. So, for most of our customers, a <u>$600 repair</u> for these electronic components seems like a large amount, especially when <u>they</u> cannot even visualize what a wire harness or printed circuit board looks like. In contrast, most <u>car owners</u> will understand and accept a $600 charge to repair the transmission on the family car.

This revised paragraph is easier to read because *car owners* and similar words are in the subject slot of most sentences. This makes the paragraph sound more focused.

In this revised paragraph, you should also notice two things. First, *car owners* are not always the exact words used in the subject slot. Synonyms and pronouns should be used to add variety to the sentences. Second, not all the sentence subjects need to be related to car owners. Later in the paragraph, for example, the term *$600 repair* is the subject of a sentence. This deviation from *car owners* is fine if most of the sentences' subjects in the paragraph are aligned. In other words, the paragraph will still sound focused, even though the subjects of one or two sentences do not align with the others.

Of course, the subjects of the paragraph could be aligned differently to stress something else in the paragraph. Here is another revision of paragraph 9 in which the paragraph focuses on repairs.

9b. <u>Repairs</u> to electronic components often lead car owners, who lack knowledge about electronics, to doubt the honesty of car mechanics. The <u>nature and scope of these repairs</u> are usually beyond the understanding of most non-mechanics, unlike the typical mechanical repairs with which customers are more familiar. For instance, the <u>importance of repairing</u> the transmission in a car is readily apparent to most car owners, but <u>adjustments to electronic components</u> like wire harnesses and printed circuit boards are not familiar to most customers—even though these electronic parts are crucial in regulating their car's fuel consumption and performance. So, <u>an electrical repair</u>, which can cost $600, seems excessive, especially when <u>the work</u> cannot be visualized by the customer. In contrast, <u>a $600 charge to repair</u> the family car's transmission is better understood and more acceptable to customers.

In this paragraph, the subjects are aligned around words associated with *repairs*. However, you will notice that paragraph 9a is usually easier for most people to read than paragraph 9b. Paragraph 9a is more readable because it has people as "doers" in the subject slots throughout the paragraph. In paragraph 9a, the car owners are active subjects, while in paragraph 9b the car repairs are inactive subjects. Much like sentences, the best subjects in a paragraph are people or things that are doing something.

The Given/New Method

Another way to write plain paragraphs is to use the given/new method to weave sentences together. Developed by Susan Haviland and Herbert Clark in 1974, the given/new method is based on the observation that readers will try to fit the new information in a sentence into what they already know. So, every sentence in a paragraph should begin with something the readers already know (i.e., the given) and end with something new that the readers do not know (i.e., the new). To illustrate, consider these two paragraphs:

10a. New Mexico is a beautiful place. Artists sometimes strike off into the mountains with sketchpads and cameras. Studios around the state are another favorite place to work. The southwestern landscapes are wonderful in this enchanted state.

10b. New Mexico is a beautiful place for artists to work. Some artists strike off into the mountains with sketch pads or cameras. Other artists enjoy working in the many studios around the state. The mountains and the studios offer places to paint the wonderful southwestern landscapes of this enchanted state.

These examples are both readable, but paragraph 10b is easier to read because each new sentence carries something over from the previous sentence.

In most sentences, the given information should appear early in the sentence, and the new information should appear later in the sentence. The given information will provide a familiar anchor or context for the readers. Later in the sentence, the new information builds on that familiar ground. Consider this longer paragraph:

11. Recently, an art gallery exhibited the mysterious paintings of Irwin Fleminger, a modernist artist whose vast Mars-like landscapes contain cryptic human artifacts. One of Fleminger's paintings attracted the attention of some young schoolchildren who happened to be walking by. At first, the children laughed, pointing out some of the strange artifacts in the painting. Soon, though, the artifacts in the painting drew the students into a critical awareness of the painting, and they began to ask their bewildered teacher what the artifacts meant. Mysterious and beautiful, Fleminger's paintings have this effect on many people, not just schoolchildren.

In this paragraph, the beginning of each sentence provides something given, usually an idea, word, or phrase drawn from the previous sentence. Then, the latter part of each sentence adds something new to that given information. By chaining together the given and new information, the paragraph gradually builds the readers' understanding, adding more information with each sentence.

In some cases, however, the previous sentence does not offer a suitable subject for the following sentence. In these situations, a transitional phrase can be used to provide the readers given information at the beginning of the sentence:

12. This public relations effort will strengthen Gentec's relationship with community leaders. With this better relationship in place, the project details can be negotiated with terms that are fair to both parties.

Here, the given information in the second sentence appears in the transitional phrase, not the sentence's subject. Transitional phrases are a good place to include given information when the subject cannot be drawn from the previous sentence.

To sum up, two primary methods are available for writing plain paragraphs: (1) aligning the subjects of the sentences and (2) using the given/new method to weave the sentences together. Both methods are helpful in proposal writing and can be used interchangeably to build more readable paragraphs. Both methods can be used as the writer weaves sentences into a plain paragraph.

When Is It Appropriate to Use Passive Voice?

Before discussing the elements of persuasive style, we should expose one boogie monster as a fraud. Since childhood, you have probably been warned against using passive voice. Teachers or supervisors may have told you that passive voice was off-limits, period. "It's bad for you," they said. "Don't use it!"

Passive voice *can* be problematic because it removes the doer from the sentence. For example, consider this passive sentence and its active counterpart:

13a. The door was closed to ensure privacy. (passive)

13b. Frank Roberts closed the door to ensure privacy. (active)

Written in passive voice, sentence 13a lacks a doer, which means it's less clear than 13b. The subject of sentence 13a, the door, is being acted upon, but it's not doing anything.

Despite your teachers' dire warnings, passive voice has a place in proposals, especially in highly technical or scientific proposals. Either of the following conditions makes a passive sentence appropriate:

- The readers do not need to know who or what is doing something in the sentence.
- The subject of the sentence is what the sentence is about.

For example, in Sentence 13a, the person who closed the door might be unknown or irrelevant to the readers. As the readers, do we need to know that Frank Roberts closed the door? Or, do we simply need to know that the door was closed? If the door is what the sentence is about and who closed the door is not relevant, then the passive voice is fine.

Consider these other examples of passive sentences:

14a. The shuttle bus will be driven to local care facilities to provide seniors with shopping opportunities. (passive)

14b. Jane Chavez will drive the shuttle bus to local care facilities to provide seniors with shopping opportunities. (active)

15a. The telescope was relocated to the Orion system to observe a newly discovered nebula. (passive)

15b. Our graduate assistant, Mary Stewart, relocated the telescope to the Orion system to observe a newly discovered nebula. (active)

In these sentences, the passive version of the sentences would be more appropriate, unless there is a reason Jane Chavez or Mary Stewart needs to be singled out for special consideration.

Also, passive sentences can help you align the subjects and use given/new strategies when developing a plain paragraph. For example, consider the following two paragraphs:

16a. The merger between Brown and Smith will be completed by May 2024. Initially, Smith's key managers will be moved into Brown's headquarters. Then, other Smith employees will then be gradually added to the Brown hierarchy to minimize redundancies. During the merger process, employees at both companies will be offered all possible accommodations to help them through the uncertain times created by the merger.

16b. Brown and Smith will complete a merger in May 2024. Initially, Bill's Trucking Service will move the offices of key managers at Smith into Brown's headquarters. Brown's human resources manager will then gradually move Smith's other employees into the Brown hierarchy to minimize redundancies. During the merger, vice presidents, human resources agents, and managers at all levels will offer accommodations to employees at both companies to help them through the uncertain times created by the merger.

Most people would find Paragraph 16a more readable because it uses the passive voice to align the subjects of the sentences on the *employees*. Paragraph 16b is harder to read because it includes irrelevant doers, like Bill's Trucking Service, and it keeps changing the subjects of the sentences. The changing subjects of the sentences make the paragraph feel bumpy to read.

In scientific and technical proposals, the passive voice is often the norm because *who* will be doing *what* is not always known or predictable. For example, in sentence 15b, the scientists may not know whether Mary Stewart will be the person operating the telescope on a given evening. More than likely, all they can confidently say is that *someone* at the observatory will be operating the telescope on a particular day. So, the passive voice is used because *who* will move the telescope is not important. On the other hand, the fact that the telescope will be moved *is* important.

When properly used, passive voice can be a helpful tool in your efforts to write plain sentences and paragraphs.

Persuasive Style

Persuasive style can help motivate readers to say yes to your ideas. Designed to motivate readers to make a decision or take action, persuasive style is best used when trying to emphasize or amplify specific ideas. As a rule of thumb, persuasive style is best used in places where you expect your readers to make decisions.

In proposals and grants, four stylistic persuasion strategies will give your writing more impact:

- Elevating the tone
- Using similes and analogies
- Using metaphors
- Changing the pace

Let's consider each of these strategies separately.

Elevating the Tone

Tone is the sound or pitch that the readers will "hear" while reviewing your proposal. Most people read proposals silently to themselves, but all readers have an inner voice that vocalizes the words as they move from sentence to sentence. By paying attention to tone, you can influence the readers' inner voice to read the proposal with a specific emotion or attitude. You can also use tone to establish a sense of authority that builds your, your team's, and your company's credibility. Tone puts a human face on the text by appealing to the readers on an emotional and authoritative level.

Writers often elevate the tone at key moments in the proposal, especially in the introduction, section openings and closings, and explanations of costs and benefits in the conclusion. For instance, you may write most of the proposal in plain style. Then, as you near the closing paragraph of larger sections or the conclusion of the proposal, you should gradually elevate the tone of your writing to increase the emotion and sense of authority.

Effective public speakers use this tone elevation technique all the time. Their speech may start with a good amount of energy, but it soon settles into a plain style. Then, when speakers near important points or their conclusions, they elevate the tone to heighten the intensity. When they hear this elevated tone, the audience knows that the speaker is nearing an important point, so they listen more closely.

One easy way to elevate tone in written texts is to decide what emotions or authority you want the readers to experience. Then, map out those feelings on a piece of paper. For example, imagine you want to establish an exciting tone as the readers begin reviewing the proposal. Put the word *excitement* in the middle of a sheet of paper or your screen. Then, as shown in Figure 10.4, map out the feelings associated with this emotion.

You can then weave these words into your proposal or grant at strategic moments. Subconsciously, your readers will detect this excited tone in your work, and their inner voice will begin reinforcing the sense of excitement you are trying to convey.

Similarly, if you want to add a tone of authority, you can map out the words associated with that tone. For instance, let's say you want your proposal to convey a feeling of safety. A map around the word *security* might look like Figure 10.5. If you weave these words associated with security into strategic places in your proposal, your readers will perceive the sense of security you are trying to convey.

Of course, writers can overdo using emotional and authoritative tones. To avoid this problem, choose one emotion and one authoritative tone for the proposal. If you use multiple emotional or authoritative tones, your readers will feel them working against each other. Also, use words from your maps sparingly. Like adding spices to food, you want to avoid over-seasoning your proposal.

Figure 10.4: Using a Concept Map to Create an Emotional Tone

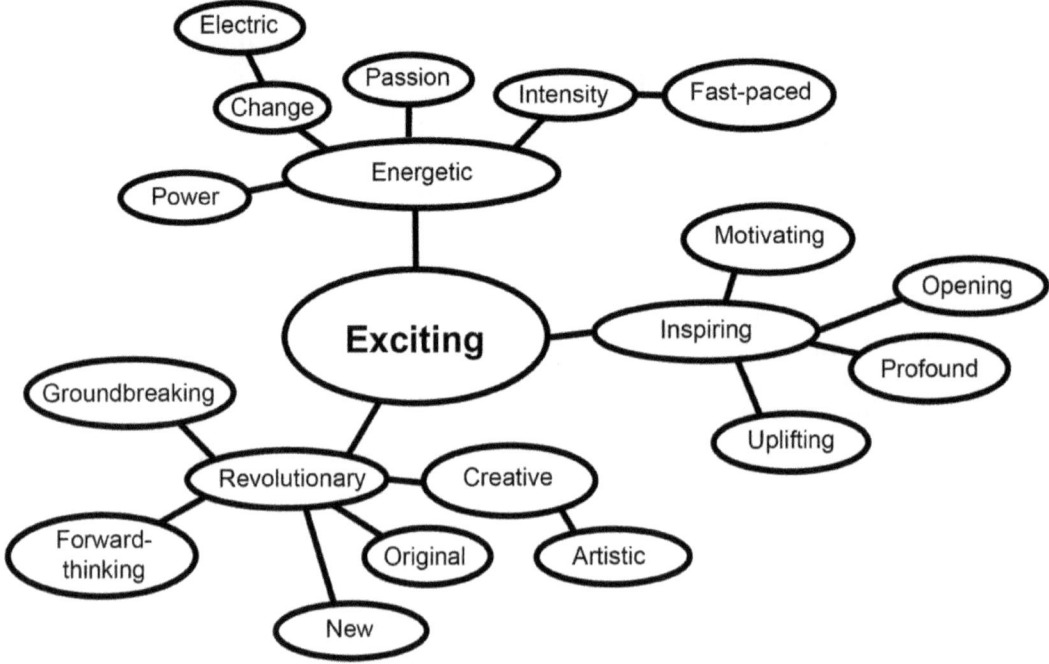

Figure 10.5: Using a Concept Map to Create an Authoritative Tone

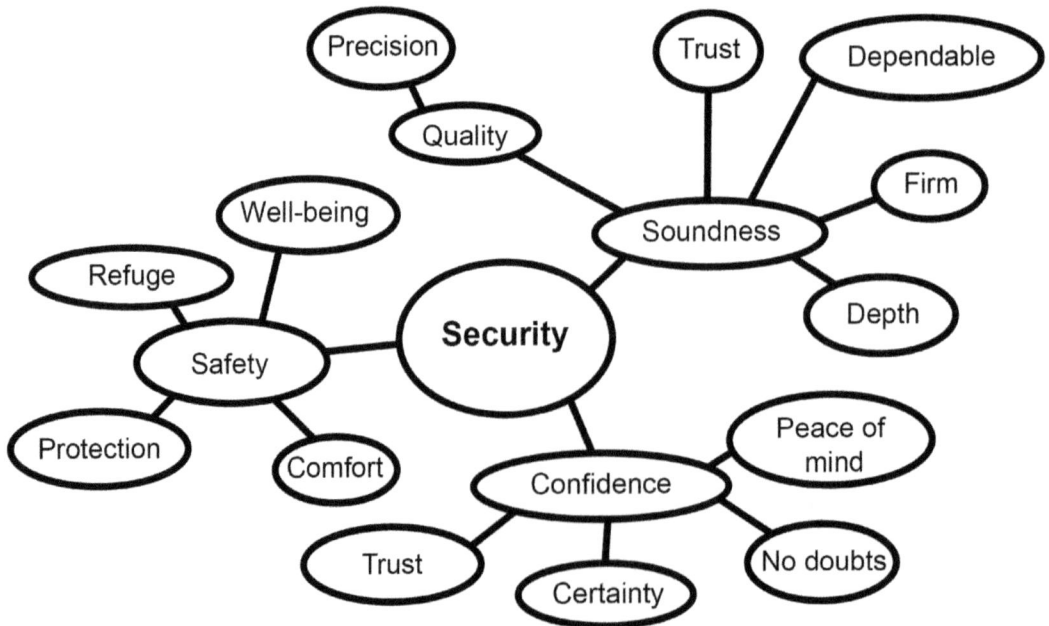

Using Similes and Analogies

Similes and analogies are rhetorical devices that help writers define difficult concepts by comparing them to simpler things. For example, let's say a proposal we are writing needs to describe a *semiconductor* to people who know almost nothing about semiconductors. A simile could be used to describe the wafer this way: "A semiconductor is like a miniature Manhattan Island crammed on a silicon disk that is only inches wide and the height of a thick piece of paper." In this case, the simile ("X is like Y") describes the semiconductor in familiar terms while creating a visual image that helps the readers understand the complexity of the semiconductor wafer.

Analogies are also a good way to help your readers visualize complex concepts. An analogy follows the structure "A is to B as X is to Y." For example, a medical analogy might be, "Like police keeping order in a city, your white blood cells patrol your body to control viruses." In this case, both parts of the analogy are working in parallel. *Police* are comparable to *white blood cells,* and *patrol* is similar to *controlling viruses.*

Similes and analogies are primarily used to provide the readers with a quick, visual understanding of something unfamiliar by comparing it to something they know about. A good rule of thumb is to use similes and analogies when the readers are less experienced with your topic. Reduce the use of similes and analogies if the readers are experts. Hyper-technical readers may find excessive similes or analogies too simplistic and inaccurate, damaging your credibility in the proposal or grant.

Using Metaphors

Though comparable to similes and analogies, metaphors work at a deeper level in a proposal or grant. A metaphor creates or reinforces a particular perspective you want your readers to adopt toward the topic or your proposed project.

Conceptual Metaphors

Conceptual metaphors are already familiar to your readers, making them particularly useful (e.g., Time is money; Life is a journey; Knowledge is light). For example, a conceptual metaphor in Western medicine is the "war on cancer." If you were writing a grant proposal to request funding for cancer research, you might weave this metaphor into your proposal. You could talk about "battles with cancer cells," "new weapons against cancer," and "the front line of cancer research." By employing versions of this metaphor throughout the grant proposal, you could reinforce a particular perspective about cancer research.

A metaphor such as "war on cancer" adds a sense of urgency to the proposal because it suggests that cancer is an enemy that must be defeated at almost any cost. Of course, cancer research is not really a war. And yet, we accept this metaphor with little question or dispute.

Conceptual metaphors are powerful stylistic tools in proposal writing because they tend to work at a subconscious level. In other words, the "war on cancer" metaphor will seem true to the readers and create a visual image in their minds. A metaphor like this one can be used in key places throughout the proposal to shape the readers' point of view, turning cancer into an enemy in their minds.

Invented Metaphors

But what if a conceptual metaphor, like the "war on cancer," is inappropriate for your proposal? In these cases, you can invent a new metaphor and then use it to create a new perspective for the readers. For example, perhaps we want our readers to view cancer as something to be *managed*, not fought. Our new metaphor, *managing cancer*, would allow us to talk about "setting objectives," "forming a research team," "listening to your cancer," or "using chemotherapy drugs as tools." We might speak of patients as managers who "set their own goals" and "strive to reach benchmarks." Doctors might become "consultants" who advise patients on managing their illnesses. This new metaphor creates an entirely different perspective than the war metaphor. It shifts the readers' perspective, urging them to think differently about how patients will handle their illnesses.

Changing the Pace

You can also regulate the readers' pace as they read through your proposal or grant. Shorter sentences tend to speed up their reading pace, while longer sentences tend to slow it down. By shortening and lengthening your sentences, you can increase or decrease the intensity of your writing. Try it. You'll see.

For example, let's say you believe a problem is urgent and must be handled immediately. The best way to increase the intensity of your proposal would be to use short sentences while you describe that problem. As the pace increases, the readers will naturally feel the urgency of the problem because they will sense the problem is getting worse. On the other hand, if you want the readers to be cautious and deliberate, you can use longer sentences to decrease the intensity of the proposal, giving the readers the impression that there is no need to rush.

Sentence length is a great way to establish the intensity you want without saying something such as "This opportunity is slipping away!" or "We need to take action now!" Longer sentences can soothe anxious readers by slowing down how quickly they are reading.

Chapter Summary and Looking Ahead

Some people may wonder about the ethics of using stylistic devices to influence readers. Most people would not question using plain sentences and paragraphs to help the readers understand the ideas in a proposal or grant. Still, using persuasive style might sound a bit like manipulating the readers.

You are manipulating the reader when you are trying to persuade them. That's what proposals and grants do. The challenge is to match your proposal's style to the readers' needs. Plain style is best for instructing the readers, giving them the facts in a straightforward way. Persuasive style is used to motivate the readers to take action.

Motivating your readers is ethical if you urge them to do what is best for both sides. When persuasive style is used properly, it matches the tone, choice of words, and pace to the problem the readers are trying to solve. It stresses important ideas at important points in the Project Plan section and the Costs and Benefits. When attention to style is coupled with the page design techniques covered in the next chapter, you can produce a compelling proposal that sounds persuasive and looks great.

Try This Out!

1. Using the six steps outlined in Figure 10.2, revise the following sentences to make them more readable:
    a. According to our survey that we conducted last Friday after the president gave his speech on crime on campus, the collection of data offers a demonstration of how important this issue is to women of college age.
    b. The meeting over the project for Guilford simply gave us confirmation that we may find it necessary to pursue the hiring of an engineer who can program in the computer languages that are more popular in applications that are reliant on AI.
    c. Due to concerns about the large number of flammable items in close proximity to the location of the fire, it was necessary that an investigation of the blaze in the southwest corner of the building be conducted by an inspector from the fire department.
2. Find three sentences that are hard to read. Use the six steps outlined in Figure 10.2 to make these sentences easier to read. Then, in a memo to your instructor, describe the steps you followed to improve the sentences' readability.
3. Use subject alignment and given/new techniques to make the following paragraph more readable:

    Because most readers equate clear writing with sincerity and trustworthiness, style is vital in proposal writing. The readers will question the soundness and honesty of an argument if the meaning of the proposal is hard to understand. After all, they wonder, perhaps the argument in the proposal is not clear to the writers if their ideas cannot be expressed in plain language. Even worse, important assumptions or facts might be hidden in the twists and turns of the sentences and paragraphs, they might suspect. On the other hand, the writers show the readers that they know what is needed and how to provide it when they submit a plainly written proposal. In my experience, clients are more easily won over by writers who submit a plain proposal rather than a proposal that is hard to read.

4. On campus or in your workplace, find a couple of paragraphs that seem difficult to read. First, align the subjects in each paragraph to see if this technique improves readability. Then, use given/new strategies to revise each paragraph. Which method works better? Would a combination of subject alignment and given/new be more appropriate in some cases?

5. Study the style of a proposal or grant found on the Internet or in your workplace. Answer the following questions:
   a. What is the tone used in the proposal?
   b. Does the proposal use similes or analogies to help the readers visualize complex concepts?
   c. Can you locate any metaphors/themes that are woven into the text?
   d. How might the persuasive style techniques discussed in this chapter improve this proposal?

6. A common metaphor in our society is the "war on drugs." With a team, think up some common phrases that demonstrate this metaphor in use. Consider the following questions:
   a. What does this metaphor imply about people who use and sell drugs?
   b. What police tactics does the metaphor suggest?
   c. What does the metaphor imply about our interactions with countries that export illegal drugs?
   d. If we wanted to handle the drug problem from a different perspective, what might be some other appropriate metaphors?
   e. How might these metaphors imply different approaches to illegal drugs?

7. While revising some of your writing (perhaps your Background section or Project Plan sections), create a specific tone by mapping out an emotion or sense of authority that you would like the text to reflect. Weave a few concepts from your map into your text. At what point is the tone too strong? At what point is the tone just right?

8. Revise by shortening all the sentences using the text from Exercise 7 above. How do these shorter sentences change the pace of your writing? Now, revise the text by lengthening the sentences. How do these longer sentences change the pace of your writing?

Case Study: The Carbon Neutral Campus Project—What Should Be the Proposal's Style?

While George worked on the budget, Anne began copyediting the Carbon Neutral Campus Project grant proposal. She was especially interested in making the proposal easier to read and more persuasive.

Anne liked editing documents more than drafting them. Writing that first draft always seemed difficult, but once she had her ideas written down, she enjoyed wordsmithing sentences and paragraphs to make them clearer and easier to read.

As she looked over the rough draft, Anne noticed that some sentences were hard to read because they didn't have obvious subjects, and many didn't use action verbs. Also, many paragraphs didn't have topic sentences and didn't flow well. Anne knew these kinds of stylistic problems were typical for a first draft.

In particular, one paragraph from the Background section caught her interest. The paragraph originally read:

> So, the legacy of Old Betsy lives on. Today's plant, the Young Power Plant (Figure 1), still burns natural gas to make steam, which is pushed through tunnels under the campus to keep buildings warm. The electricity on campus mostly comes from the Four Corners Power Plant, a massive natural gas-burning plant that is located west of Farmington, New Mexico. With the increase in electricity—using devices on campus like machinery, computers, and electric HVAC systems, the electricity needs of the campus outstripped the generating capacity of its on-campus power plant in the mid-1970s.

She noticed that each sentence in the paragraph used a different subject, and the subjects of some sentences were not always easy to identify.

Anne began revising the paragraph by asking what the paragraph was about. She decided the paragraph was about the campus's reliance on natural gas for energy. She then restructured the sentences in the paragraph to focus on the campus as a subject.

> So, the legacy of Old Betsy lives on. Presently, the Durango University campus is heated by the Young Power Plant (Figure 1), which still burns natural gas to make steam and then pushes it through tunnels under the campus. The campus draws its electricity from the Four Corners Power Plant, a massive natural gas-burning plant that is located west of Farmington, New Mexico. The campus's electricity needs outstripped the generating capacity of its own power plants in the 1970s, mainly because of new electricity-using devices on campus like machinery, computers, and electric HVAC systems.

Anne's revised paragraph was easier to read. The transition sentence (first sentence) starts the paragraph by referring to something in the previous paragraph. The second sentence is the topic sentence, providing a claim the paragraph strives to prove. The remainder of the paragraph uses facts and logical reasoning to support the topic sentence.

After Anne finished revising all the paragraphs in the proposal to make them more readable, she decided to amplify parts of the proposal with persuasive style.

First, she began thinking about the tone she wanted the proposal to convey. She found herself returning to the word "innovative" as a possible tone for the whole document. Putting the word *innovative* in the middle of a sheet of paper, she mapped out some words associated with this concept.

Figure 10.6: Setting an "Innovative" Tone

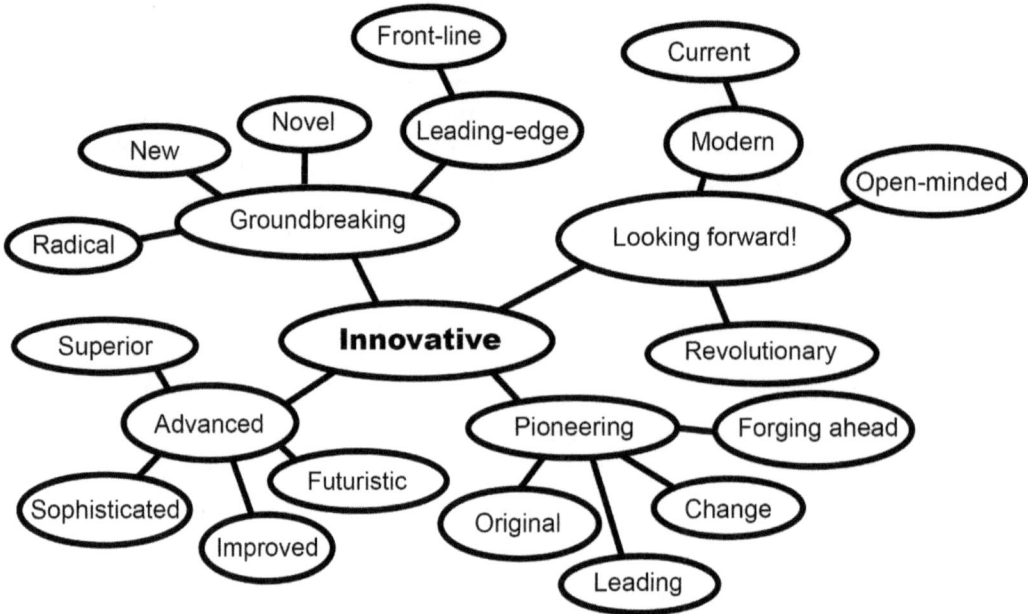

With her map finished, Anne wove some of these words into the introduction, opening paragraphs, closing paragraphs, and the Costs and Benefits section of the proposal. She wanted these words to give the readers the sense that something new and innovative was being proposed.

Then, Anne decided she needed a metaphor that would add a visual image to the grant. She began to imagine the Durango University campus as a "village" trying to become self-sustaining. She wrote, "Durango Campus is a village," and started using this metaphor to generate a new way to describe the campus. She used an AI application software to help her come up with some variations of the metaphor:

Durango Campus as a Village

- Faculty, staff, and students as neighbors
- Growing independence
- A sense of unity and cooperation
- Relying on ourselves and each other
- Many cultures living in one place
- Working together for the common good
- Looking out for each other
- Common identity despite our differences

The "village" metaphor added warmth to the proposal by drawing on words and phrases associated with a community.

When Anne finished revising it, the proposal included the same content as before. However, her use of plain style techniques to improve sentences and paragraphs made the proposal much clearer and more readable. Meanwhile, her attention to the tone and the "village" metaphor gave the proposal more depth and energy.

If you would like to see how the writing style of the Carbon Neutral Campus proposal turned out, please go to Appendix A.

11 DESIGNING PROPOSALS

"How You Say Something..."

The old saying, "How you say something is what you say," is truer now than ever when writing compelling proposals and grants. Not long ago, document design was considered a luxury in proposals, not a necessity. But today, readers expect proposals to be visually interesting and engaging. They expect a positive first impression. They expect the proposals to be more accessible so they can read more efficiently and quickly locate the proposal's most important ideas.

As a reader yourself, you probably cringe when a minimally designed document with massive blocks of text fills your screen. You know you're in for a tough slog. Your readers feel the same way. Good design allows readers to choose how *they* want to read the proposal, not how the writers think they should read it. Your readers will skim some areas of the proposal and read others closely.

Readers are *raiders* for information. They don't read proposals for pleasure, and they are looking for specific kinds of facts, analysis, and reasoning. Your proposal should be designed in a way that helps them locate the information they need to make a decision. Good design gives the readers multiple *access points*, allowing them to enter the text from various places.

Good design is more than making a proposal or grant look nice. An effective design increases the accessibility of the text by highlighting important information and allowing the readers to use the document to suit their needs. A well-designed proposal also establishes a particular tone, signaling the attitude, competence, and quality of the people submitting the proposal or grant.

Of course, the content of your proposal is still its most important feature. Even the best design can't hide a badly conceived project or a flawed understanding of the current situation. However, design can make a positive impression on the readers while making important information easy to locate. Good design gets your readers leaning into your proposal before they even read a word.

Four Principles of Design

Though professional document design takes time and practice, you can learn basic principles of design that will help you make better decisions about how your proposal or grant should look. This section will discuss four simple design principles derived from gestalt psychology, which has deeply influenced the graphic arts.

The core assumption of gestalt design is that humans do not view their surroundings passively. Instead, they instinctively look for relationships among objects and spaces, creating wholes that are more than the sum of their parts. For example, in Figure 11.1, most people will see a white square in the center with rounded corners. They might even see an X in the middle of that square. Why? The square, after all, is not really there. Gestalt design suggests that people interpreting this kind of graphic create a whole out of the parts, seeing a square and perhaps even an X where they don't exist.

The four design principles discussed in this chapter are balance, alignment, grouping, and consistency.

Figure 11.1: The Whole Is More Than the Sum of the Parts

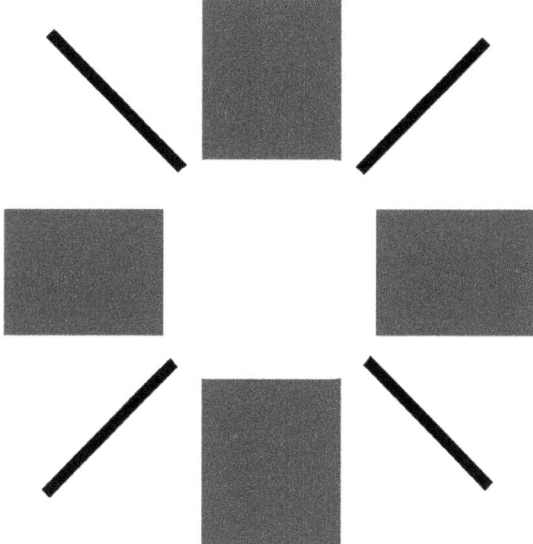

Design Principle 1: Balance

On a balanced page, the design elements offset each other to create a stable feeling in the text. Imagine a page is balanced on a point. Each time you add something to the left side, you should add something to the right side to maintain balance. Similarly, when you add something to the top of the page, you should add something to the bottom.

Figure 11.2, for instance, shows an example of a balanced and unbalanced page. The page on the left feels stable because the design elements have been balanced evenly on the page. The page on the right looks unbalanced because items on the left do not offset the items on the right side of the page. Moreover, the right page is bottom-heavy because the text is bunched up toward the bottom of the page. Readers would find the balanced page on the left more attractive and easier to read.

One thing to keep in mind is that a balanced page is not necessarily a symmetrical page. As shown in the left page in Figure 11.2, the two halves of the page do not need to mirror each other, nor do the top and bottom need to look the same. Instead, the sides of the page should offset each other to create a sense of balance.

Figure 11.2: Balanced vs. Unbalanced Page Designs

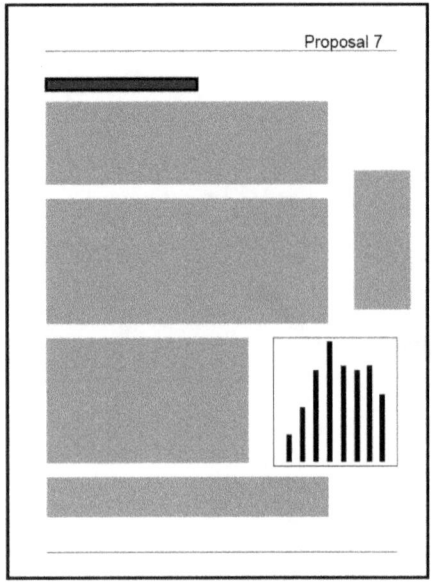 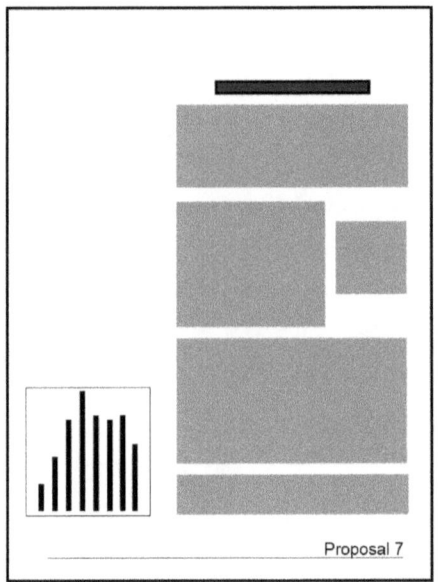

When balancing a page layout, graphic designers often talk about the *weight* of items on a page. What they mean is that some items on a page will attract the readers' eyes more than others. A color picture, for example, has more weight than printed words because readers' eyes tend to be drawn toward pictures and colors. Similarly, irregular shapes weigh more than common shapes because readers are attracted to unusual objects. Here are some general guidelines for weighting the elements on a page:

- Items on the right side of the page weigh more than items on the left.
- Items on the top of the page weigh more than items on the bottom.
- Big items weigh more than small items.
- Pictures weigh more than written text.
- Graphics weigh more than written text.
- Items with color weigh more than black and white.
- Items with borders around them weigh more than items without borders.
- Irregular shapes weigh more than regular shapes.

Here's a simple trick if you are uncertain about the weights of various page elements. Close your eyes for 15-30 seconds, then open them and look at the page. Note the order in which your eyes scanned various elements on the page or screen. More often than not, your eyes will be drawn to the "heaviest" design elements first.

When designing a model page for a proposal, create a layout that allows you to keep the text as balanced as possible.

Grids that Balance a Page Layout

A time-tested way to create a balanced page design is to use a *page grid* to balance the written text and graphics on the page. Grids divide the page vertically into two or more columns. Figure 11.3 shows a few standard grids and how they might be used.

In most cases, as shown in Figure 11.3, the columns on a grid do not translate into columns of written text. The grid is used to balance the text on the page, often allowing images to overlap one or more columns.

Figure 11.3: Using Grids to Balance Page Designs

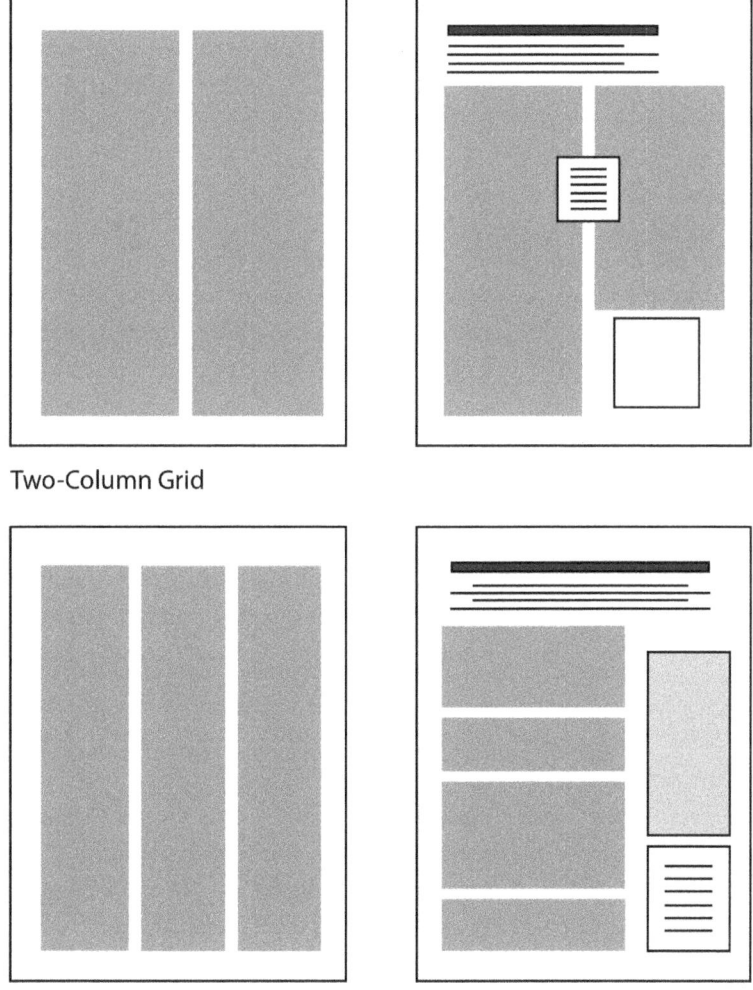

Two-Column Grid

Three-Column Grid

Why use a grid in the first place? You might be tempted to expand the margin on the right or left side in an ad hoc or makeshift way. The problem with this approach is that readers subconsciously sense the design's inconsistent structure. As gestalt design implies, readers subconsciously look for regular patterns or shapes. If no grid is used, they will try to imagine a grid anyway, making them work harder than necessary to understand the text. In the long run, a grid-based page design offers more flexibility than an ad hoc design. An ad hoc layout may work for a few pages, but as images, charts, margin text, and other graphics are added, the page design will grow increasingly difficult to manage.

Other Balance Strategies

Several other document design features can also be used to balance documents while highlighting important points and breaking up large blocks of text.

Pullouts—Quotes or paraphrases can be pulled from the body text and placed in a special text box to catch the readers' attention. Pullouts break up large blocks of text and create access points for the readers. Magazines, for example, frequently use pullouts when a picture or graphic is not available to break up a page of words. A pullout should draw its text from the page on which it appears. Often, the pullout is framed with rules or a box, and the text wraps around it (Figure 11.3).

Margin Comments—Key points or comments can be summarized in the margin of the proposal. When a grid is used to design the page, one of the margins might leave enough room to include an additional list, offer a special quote drawn from the body text, or provide a simple illustration. In a large proposal, margin comments might even remind the readers where they are in the proposal by restating its outline and highlighting the main points of the section (Figure 11.3).

Sidebars—Examples or anecdotes can be placed in sidebars to reinforce the main text. In proposals, sidebars can be used to explain a process in more detail or describe a previous project that was a success. Sidebars should never contain essential information. Instead, they offer supplemental information that enhances the readers' understanding.

Pullouts, margin comments, and sidebars can be used to balance a text and break up large blocks of words. They also enhance the readability of the proposal by reinforcing the main points and providing supplemental information.

A one-column design like the one shown in Figure 11.4 will work if necessary. In a one-column grid, graphics and text can be centered in the middle of the column. A one-column grid tends to be traditional and text-dominant while providing only limited flexibility in the placement of graphics. But it's a safe choice if you're new to document design.

Figure 11.4: Using Alignment to Show Hierarchy of Information

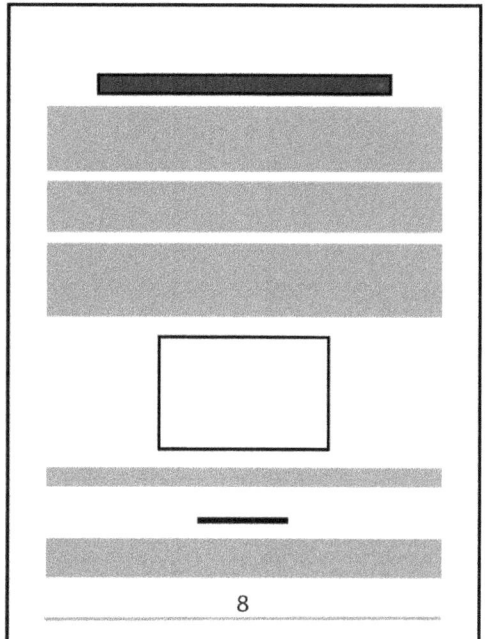

Design Principle 2: Alignment

Alignment is the use of vertical white space to help the readers identify the various levels of information in a proposal. The simplest alignment technique uses an indented list to offset a group of items from the body text. An indented list signals readers that the listed items supplement the text in the surrounding paragraphs. In Figure 11.4, for example, the page on the left gives no hint about the hierarchy of information in the text, making it difficult for a reader to scan the text. Meanwhile, the page on the right uses indented lists to signal the hierarchy of the text. The indented material is easily recognized as supplemental information.

In a proposal, blocks of text can be aligned to show the hierarchy of information. Examples or explanatory information can be indented to signal that they should be considered separately from the body text. Figure 11.4 illustrates how information can be indented to signal various levels in the text.

Essentially, alignment uses white space to create vertical lines in the text. The readers will mentally draw the vertical lines down the page, seeing aligned elements as belonging to the same level of importance.

Design Principle 3: Grouping

Grouping visually divides a page into smaller, more comprehensible parts, especially for readers skimming the document. A large block of text, perhaps a one-column page with no headings and no indentation, will feel uninviting to a reader and might be difficult to read. Grouping allows you to break up the page by providing the readers with more white space and giving them a variety of access points at which to enter the text.

Headings

Headings signal new topics to the readers while helping them see the overall organization of the proposal. With a quick scan of the headings in the document, the readers should be able to quickly identify how the information in the proposal is organized.

Figure 11.5 Levels of Headings

> # Title of Proposal
>
> ## FIRST-LEVEL HEADING
> This first-level heading is 18 pt. Helvetica, boldface with small caps, and this paragraph is 11pt. Garamond. Notice that this heading is significantly different from the second-level heading, even though both are in the same typeface. This heading is also hanging because it is placed further into the margin than the regular text. Use consistent spacing above and below each head (e.g., 24 pts. above and 6 pts. below).
>
> ## *Second-Level Heading*
> This second-level heading is 14 pt. Helvetica with italics. Usually, less space appears above and below this head (perhaps 18 pts. above and 3 pts. below).
>
> ### Third-Level Heading
> This third-level heading is 12 pt. Helvetica with no bolding. Often, no space appears between a third-level heading and the body text, as shown here.
>
> **Fourth-Level Heading.** This heading appears on the same line as the body text. It is designed to signal a new layer of information without being too prominent.

In a larger document like a proposal, headings should consistently highlight the various levels of information for the readers (Figure 11.5). A first-level heading, for example, should be sized significantly larger than a second-level heading. In some cases, first-level headings might use all capital letters (ALL CAPS), small capital letters (SMALL CAPS), or **boldface** to distinguish them from the font used in the body text.

Second-level headings should be significantly smaller and visually different than the first-level headings. Whereas the first-level headings might use all caps, the second-level headings might capitalize only the first letter of each word (excluding articles and short prepositions, e.g., "Marketing that Works with Generation Next"). First- and second-level heads are often boldfaced. Third-level headings might be italicized or placed on the same line as the body text. Figure 11.5 shows five levels of headings, illustrating how each level is visually different from the others.

Horizontal and Vertical Rules

In page design, horizontal and vertical rules are straight lines that can be used to carve the proposal into larger blocks. Rules should be used judiciously in a proposal because they can impede the reader's progress through the text. Too many rules make the document look like it has been chopped up into small bits and pieces. When used properly with headings or to set off an example, horizontal and vertical rules can help readers identify larger groups of information in a proposal.

Figure 11.6 shows how rules can be used to divide a page into parts. The horizontal and vertical rules carve the text into larger chunks, helping the readers see them as groups.

Borders

Like rules, borders are also used to group text into units. Borders, however, tend to be even more isolating than rules because they enclose text completely, setting it off from the rest of the information on a page. Borders are best used to set off examples, graphics, pullouts, and sidebars that supplement the body text. For example, in Figure 11.6, a border is used to create a sidebar in the right margin of the page.

Borders can be helpful tools for grouping information, but they should be used selectively. When borders are used too often, readers become immune to their grouping effects and skip reading the information inside.

Design Principle 4: Consistency

The final design principle, consistency, means that each page should be designed similarly to other pages in the document. Pages should follow a predictable pattern in which design features are used consistently throughout the proposal. Four techniques can be used to give each page in a proposal a consistent look: headers and footers, typefaces, lists, and labeling graphics.

Figure 11.6: Grouping with Rules and Borders

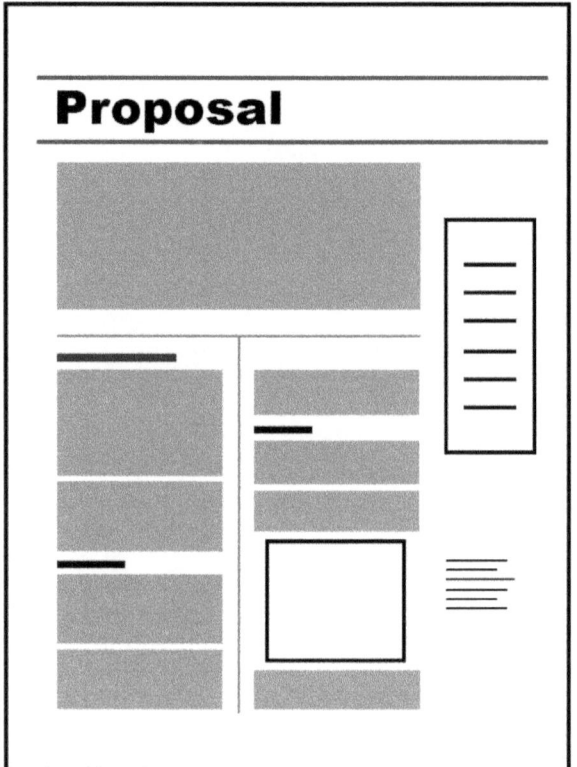

Headers and Footers

Even the simplest word-processing software can put a header or footer at the top and bottom of each page. As their names suggest, the header runs across the page's top margin (the head), and the footer runs along the bottom of the page (the foot). In many proposals, headers and footers usually include the title of the proposal and perhaps the bidding company's name. Also, page numbers should appear in the header or footer because readers need them to discuss specific parts of the proposal.

Headers and footers may include design features like a horizontal rule or company logo. These consistent features visually bring the pages together into one document.

Typefaces

As a rule of thumb, a proposal should use no more than two typefaces. Most page designers will choose two typefaces that are distinctive from each other, usually one serif typeface and one sans serif typeface. A serif typeface, like Times or Bookman, has small tips (serifs) at the end of the main strokes in the letters. A sans serif font does not include these small tips.

There are no rules for typeface usage, just guidelines. To readers in North America, serif fonts like Times or Garamond tend to look more formal and traditional, while sans serif fonts like Helvetica or Arial appear informal and modern. In North America, page designers often use a sans serif typeface in headings, headers, and footers to make their proposal look more progressive. Then, they use a serif font in the body text because North American readers usually report that they find serif fonts easier to read at length.

Some international readers, like those from the United Kingdom, are more accustomed to body text that uses a sans serif font, such as Helvetica. So, they find sans serif fonts easier to read at length.

Sequential and Nonsequential Lists

In proposals, lists are useful for showing a sequence of tasks or a group of similar items. Lists tend to fall into two basic categories: sequential and nonsequential. Sequential or numbered lists present items in a specific order. For instance, a sequential list can be used in a proposal to rank objectives or show the steps in a procedure. In sequential lists, numbers or letters are used to show the hierarchy or ordering of the items. Nonsequential lists, on the other hand, use bullets, checkmarks, or other icons to show that the items in the list are essentially equal in value.

Lists are handy tools for making information more readable, and you should look for opportunities in your proposal to use them. When you include a list, use sequential and nonsequential lists consistently. In sequential lists, numbering schemes should be the same throughout the document. For example, you might choose a numbering scheme like 1), 2), 3). If so, don't number the following list 1., 2., 3., and others A., B., C., unless you have a good reason for changing the numbering scheme.

Similarly, use the same bullets or icons when setting off lists in nonsequential lists. Do not use bullets with one list, checkmarks with another, and checkboxes with a third. These inconsistencies confuse the readers because they won't understand why various lists appear different.

To avoid these problems with lists, decide up front how your proposal will use them. Choose one format for sequential lists and another for nonsequential lists. Then, use these formats consistently throughout the proposal.

Labeling of Graphics

Graphics, such as tables, charts, pictures, and graphs, should be labeled consistently throughout the proposal. In most cases, the label above a graphic will include a number and a title (e.g., Chart 5: Forecast of Future Sales). In some cases, though, the number and title might appear below the graphic. Again, just be consistent in how you label graphics. Chapter 12 will further discuss how to label graphics in proposals.

The Process of Designing Proposals

Balance, alignment, grouping, consistency—once you know these four basic principles, designing proposals and grants becomes much more manageable. These four principles form the basis of a *design process* that you can use to design almost any text.

Using a consistent design process becomes especially important when working with a team on a proposal. Your team should agree on the design before they go off to write their sections of the proposal. They can then draft their sections in a way that fits the overall design of the proposal. Giving your team design guidelines to follow will save time later.

Here is a five-step process will help you create effective layouts for proposals:

1. Consider the rhetorical situation
2. Thumbnail a few example pages.
3. Create a design stylesheet.
4. Sketch a few generic pages.
5. Edit the design.

Step 1: Consider the Rhetorical Situation

Start the design process by revisiting your understanding of the rhetorical situation, which was explained in Chapter 3. Specifically, pay attention to the unique needs and characteristics of the primary readers and the physical contexts (e.g., office, worksite, board meeting, airplane) in which they will read and use the proposal.

Step 2: Thumbnail a Few Model Pages

With the readers and context in mind, sketch a few possible page layouts that fit the rhetorical situation. Graphic artists often start designing pages by sketching a few *thumbnails* freehand or on a computer. Thumbnails are miniature drawings of various pages. Try folding a piece of paper into 4 or 8 panels in which you can sketch sample pages. Thumbnails take a few moments to draw, but they will allow you to look over possible designs before you commit to a particular page layout—saving you time and effort.

Step 3: Create a Design Style Sheet

Record your design decisions in a style sheet. After sketching thumbnails or creating a page layout, write down some of your decisions about various design elements of the document. These features can be handled on five levels:

Line Level—font, font size, and use of italics, bolding, and underlining

Paragraph Level—spacing between lines (leading), heading typefaces and sizes, indentation, justification (right, center, left, full), sequential and nonsequential lists, column width

Page Level—columns, headers and footers, rules and borders, use of shading, placement of graphics, use of color, pullouts, sidebars, page numbers, use of logos or icons

Graphics Level—captions, labeling, borders on graphics, use of color, fonts used in tables, charts, and graphs

Document Level—binding, cover stock, paper size, color, weight, and type (glossy, semi-gloss, standard), section divider

Making all these design decisions up front can be daunting, but making these decisions earlier than later will save you time. Ask each team member to follow the style sheet as closely as possible. That way, the final draft will require less design-related revision and editing when the proposal comes together.

Step 4: Create a Template

With your thumbnails sketched and a style sheet created, use some or all the proposals' content to create a template that each page in the proposal will follow. As you add content to your design, you will likely discover that your template needs to be modified to keep the design functional and consistent. Mark these modifications in your style sheet.

Step 5: Edit the Design

Editing is an essential aspect of design. After you have completed designing the pages and adding the written text and graphics, you should spend time specifically revising and editing the proposal's design. To help you edit, look back at the rhetorical situation, your thumbnails, and your style sheet. Does the final design fit the readers and the context in which they will read the proposal? Does the final design reflect the visual qualities you wanted as you sketched your thumbnails? Are there any places where the final design needs to be revised to fit the style sheet? Or does the style sheet need to be modified to fit your design decisions? Ultimately, make sure the final, edited design fits the proposal's rhetorical situation.

Chapter Summary and Looking Ahead

Sometimes, proposal writers mistakenly view document design as a luxury, not a necessity. The content, not the design, is what readers care about, right?

Not true. Always remember that "how you say something is what you say." How a proposal looks says a significant amount about how your company or organization does business. In the long run, spending the few hours necessary to design an attractive, functional proposal will be worth the effort. The next chapter will show you how to create and use graphics, another important design element in proposals and grants.

Try This Out!

1. Choose three full-page advertisements from a magazine or a website. How did the designers use balance, alignment, grouping, and consistency to design these advertisements? How is the rhetorical situation (topic, purpose, readers, context of use) reflected in the design of these advertisements?

2. Analyze the design used in an existing proposal or grant. Write an email to your instructor in which you discuss how the proposal's design uses the principles of balance, alignment, grouping, and consistency. Then, note where the proposal strays from these principles. Based on your observations, do you think the proposal's design is effective? How might it be improved?

3. For a practice or real proposal of your own, go through the five-step design process discussed in this chapter. Study the rhetorical situation from a design perspective, then thumbnail a few pages for your proposal. Record your decisions in a style sheet. Then, fold a piece of paper into 4 or 8 panels and thumbnail a few generic pages for your proposal.

4. Study the page layouts used in other kinds of documents (e.g., newsletters, posters, or books). Can any of these designs be adapted as models for proposals? Why might you try different designs that break away from the more traditional designs used in proposals?

5. Study the designs of the example proposals in Appendix A of this book. How do these document designs change the tone and readability of these documents?

Case Study: The Carbon Neutral Campus Project—How Should the Proposal Be Designed?

Calvin was already thinking about the proposal's design before the team was writing the first draft. As a contractor, he knew that appearances were always important. Designing a building, for example, does more than just make it look nice. A well-designed building is more functional and usable, and it also makes a statement about the quality of the company or organization that works inside.

Calvin knew that the same was true with proposals. Sure, a colorful, well-designed proposal gets the readers' attention, but good design can also make it easier to read and use. He always wanted his proposals bidding for projects to be both attractive and accessible. That way, the readers would want to spend time with his proposals, and they could find the information needed to make decisions.

Calvin also appreciated the well-designed proposals he received from his subcontractors. He was very busy and didn't have time to fight through badly designed proposals with dense text blocks and inconsistent formatting. Proposals that used balanced pages, columns, headers and footers, captions, labels, graphics, and color were more attractive and easier to read.

Calvin texted Tim, who was busy creating graphs, tables, photos, and other visuals for the Carbon Neutral Campus proposal. The two of them agreed that they should come up with a consistent design for the proposal.

They began designing by thinking about the rhetorical situation. "Let's look at our readers. What design would most appeal to them?"

"Well, the primary readers are the board members at the Tempest Foundation," said Tim, "but we have quite a few other readers to consider too, including President Wilson, the Board of Regents, students, faculty, staff, and the local community."

"I don't know how we're supposed to design a document for all these different people," Calvin said. "President Wilson probably wants something more formal and academic, but the folks in our community aren't going to want to read something that looks too academic or complicated."

Tim agreed. "OK, maybe we should concentrate on our primary readers, the Tempest Foundation Board. Above all, we need to design the document so they can quickly locate the information they need. What are their needs, and how do their busy schedules affect our design decisions?"

They looked over their notes about the reviewers in their rhetorical situation worksheets. The Tempest Board seemed to be made up mostly of business professionals who were volunteering their time. They would only have limited time and energy to read a stack of proposals.

Calvin pointed out that the readers' physical context would be very important to the design of the grant proposal. The board would probably discuss the grant at a meeting, so important facts, plans, and other details needed to be easy to locate.

The proposal's design would need to accommodate readers who don't have the time or energy to read a proposal all the way through. They would probably skim most of it. They would set it down and pick it up later. So, obvious access points to start reading again would be important.

With these readers and contexts in mind, Calvin thumbnailed a few possible pages for the proposal (Figure 11.7).

"I like that," Calvin responded. He drew boxes for pull quotes on each page.

They decided that the first page of the proposal would double as the cover page. That way, the readers would be put right into the proposal itself.

Figure 11.7: Thumbnail Sketches of Possible Page Designs

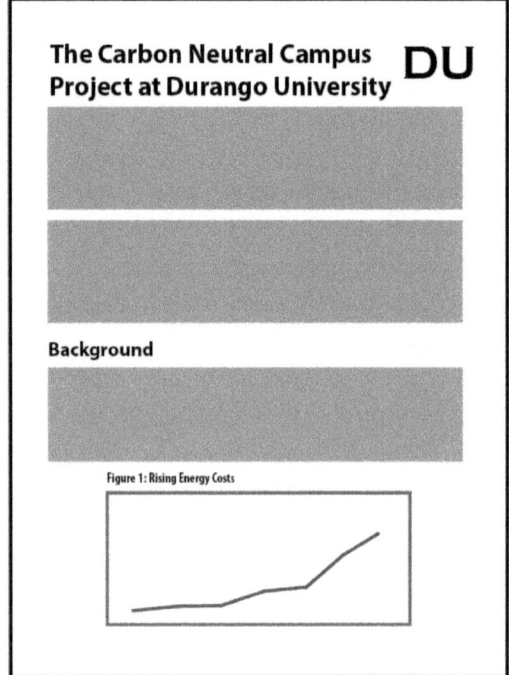
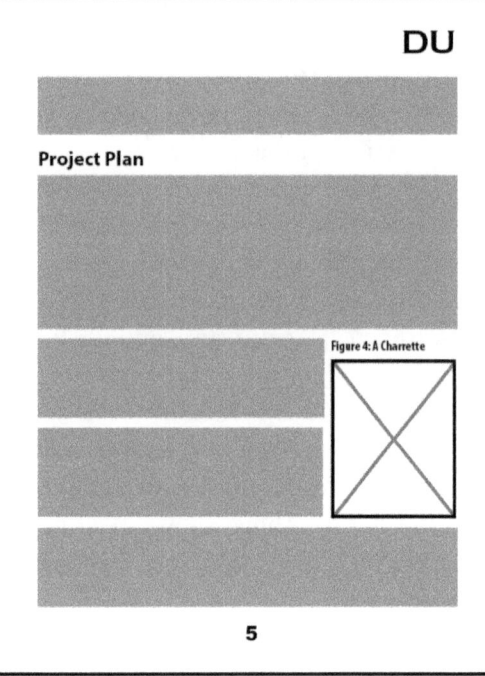

Looking at Calvin's thumbnails, Tim said, "I like how you did that. Avoiding a formal cover page makes the proposal look more inviting and not as stodgy. That large title on the first page draws the readers into the text. A picture on the front page would also grab the readers' attention."

"Good idea," Calvin replied. He integrated Tim's suggestions into the thumbnails.

Tim needed to go to class, so Calvin rode his bike back to his office to finish his thumbnail sketches. At his computer, he began typing up a simple style sheet that would guide the grant-writing team's decisions about design. He found a basic template on the Internet that he could modify.

The style sheet and template would be especially helpful as they revised and put the whole proposal together. They would help the team create a consistent design.

From prior experience, Calvin knew his decisions in the style sheet would need to be modified as the proposal evolved. But, creating a design style sheet only took about five minutes, and it would give the group some basic design guidelines to work from. Creating a template took about a half hour.

Calvin tried filling the template with some text from the first draft of the proposal. He revised the template in places where the page design didn't work with the written text or graphics.

As he designed the pages, it became obvious that he would need to collect more information before finalizing how the proposal would look. First, he wanted to add some pullouts that featured quotes from President Wilson, the climate crisis experts, and community members. Second, the proposal needed photographs, charts, and tables for the body pages. Tim was collecting those graphics, so Calvin left borders where he thought images, graphs, and tables might appear.

Calvin would share his designs at the next meeting of the Carbon Neutral Campus team. He knew changes would need to be made to the design, but he felt pretty good about his and Tim's design decisions.

If you would like to see how the design of the Carbon Neutral Campus proposal turned out, please go to Appendix A.

12 USING GRAPHICS

The Need for Graphics

Given the ease with which today's computers and artificial intelligence applications (AI) create graphics, you should include visuals in your proposals and grants. Your readers will expect to see graphs, charts, tables, and photographs in the document. We live in an increasingly visual society, so your readers will often rely on what they see in your proposal as much as what they read.

Graphics do more than offer images and visualize data in your proposal or grant—they help you tell stories that will be persuasive to your readers. There are other benefits to graphics:

- Graphics can keep readers from simply scanning the proposal. When readers come across an interesting visual, they typically look into the written text to find an explanation for the displayed data. Your proposal's graphics serve as *access points* where readers can re-enter the written text.
- Graphics break up large blocks of written text, providing the readers with resting places in the document. A proposal that forces readers to march through pages and pages of written text is difficult to read. Visuals give your readers opportunities to pause and consider the ideas in your proposal
- Graphics reinforce your argument visually in the proposal. In our visual age, readers tend to trust what they can see. A well-placed graphic bolsters the proposal's argument by showing, not just telling.

This chapter teaches you how to use graphics effectively in your proposals and grants. We will begin by reviewing basic guidelines for using graphics in proposals and grants and then describe some commonly used graphs and charts.

Guidelines for Using Graphics

Graphics should reinforce and clarify the message of the written text. Use properly, they can clarify the details and numbers to create powerful images in the readers' minds. They can also reveal or highlight special relationships among data points, organizations, and people.

There are four general guidelines for using graphics in a proposal or grant. A graphic should do the following:

Tell a simple story—At a glance, your readers should be able to grasp the story the graphic is telling. For example, a rising line in a graph signals that something is going up. Keep your graphics as simple as possible, so readers don't need to work too hard to figure out the story each tells.

Reinforce the written text, not replace it—Graphs and charts should be used to support the written text, not replace it. When referring the readers to a graph, chart, or image, you should also explain what it shows.

Be used ethically—There are many ways to distort or misrepresent data in graphs and charts. Likewise, photo editing, especially with artificial intelligence (AI) applications, can express false or dishonest ideas.

Be labeled and placed properly—Graphics should be numbered and titled unless they are being used solely to decorate the text. In the written text, you should refer to the graphic by number (e.g., Table 2.3). A graphic should appear on the same page or after it is referred to in the text. That way, your readers don't need to double back to find the graphic when you explain it.

These guidelines will help you decide whether a graphic is doing what you intend it to do. Each time you consider adding a graph, chart, illustration, or photo to your proposal, ask yourself whether it a) tells a simple story, b) supports the written text, c) shows the truth, and d) is labeled and placed properly. If you can answer yes to these four questions, the graphic will be helpful to the readers.

Using Graphics to Display Information and Data

A variety of graphics are available for displaying information and data. Each type of graphic allows you to tell a different story with the information you are presenting.

Line Graphs

Line graphs in proposals and grants are usually used to show trends over time. In a typical line graph, the dependent variable axis (y-axis) displays a fluctuating quantity like income, sales, production, and so on. The independent variable axis (x-axis) is divided into consistent increments like years, months, days, or hours.

For example, Figure 12.1 shows a line graph that plots the population changes (dependent x-axis) in mountain lions and bighorn sheep over a five-year period (independent y-axis). As you can see, the line graph allows the readers to see the story of how these two animal populations are interrelated.

Figure 12.1: A Line Graph That Shows a Trend

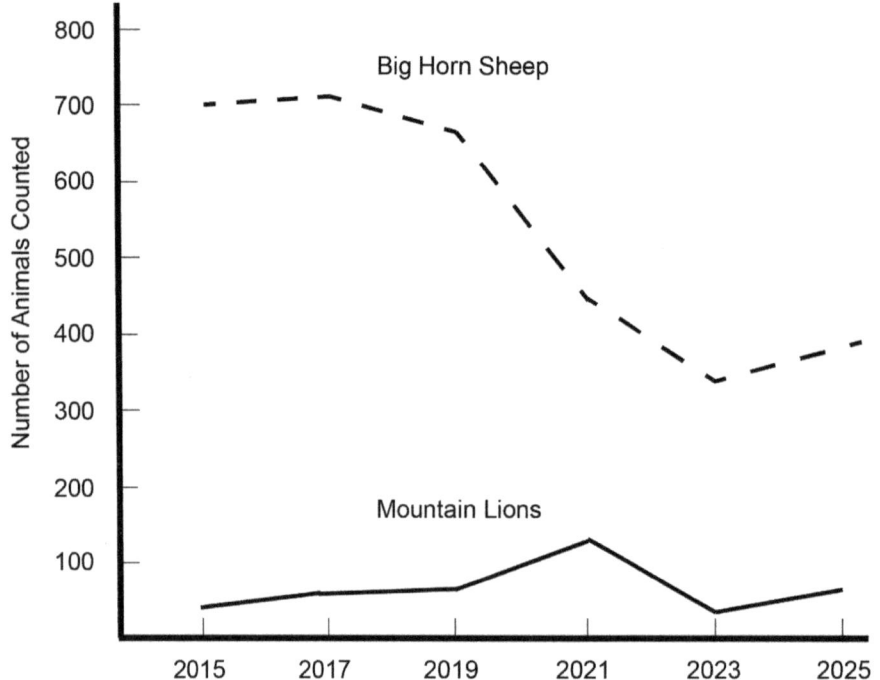

Figure 17.2: Relationship Between Populations of Big Horn Sheep and Mountain Lions in the Luna Mountain Range

The drawback of line graphs is that they don't provide the readers with exact figures. In Figure 12.1, can you tell exactly how many big horn sheep were counted in 2022? Not really. Line graphs are strongest when the trend is more important than the exact figures behind that trend. If exact figures are needed, you can still use a line graph, but provide your readers with the data in a table in an appendix.

Bar Charts

Like line graphs, bar charts can show trends over time. Bar charts are especially well suited to showing changes in volume over time. For example, Figure 12.2 illustrates how the percentage of patients (a volume) who had H1N1 Flu fluctuated by age group.

The advantage of a bar chart over a line graph is that the columns visually reflect a physical quantity. The columns' sizes allow readers to make easier comparisons among data points because some columns are visibly larger than others. A line graph, in contrast, only plots vertical points without providing a complete sense of the volume.

Columns that represent dependent variables in a bar chart (usually on the y-axis) should always start at zero. If the y-axis does not start at zero, the differences among the columns will be improperly exaggerated, giving the readers the impression that the differences among the columns are more significant than they are. In almost all bar charts, it is unethical to start the independent variable axis (again, usually the y-axis) at a number other than zero.

Figure 12.2: A Bar Chart that Illustrates a Volume

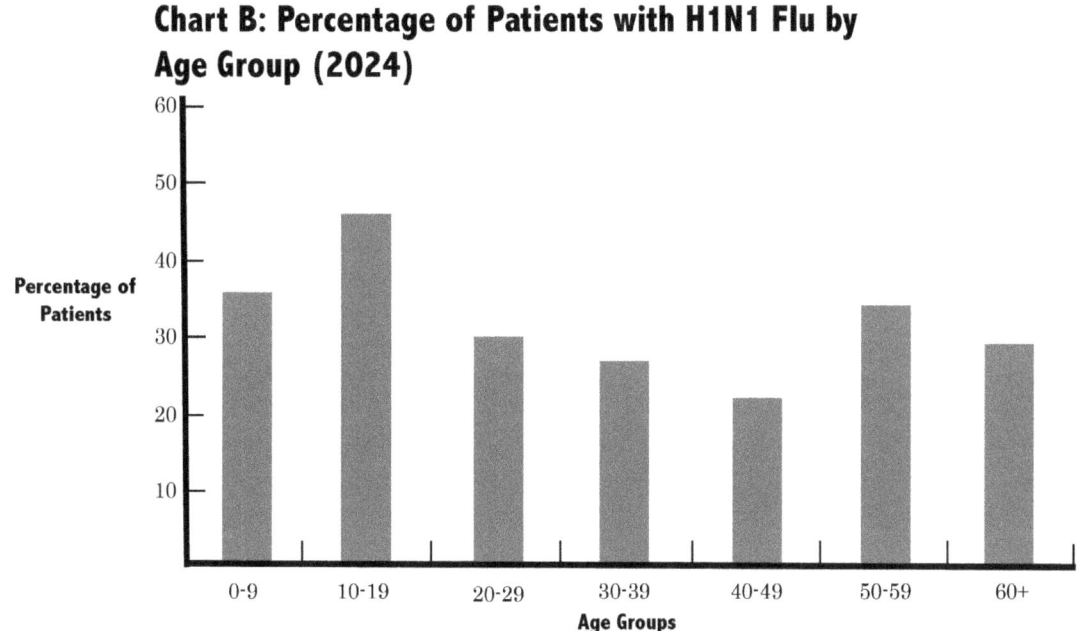

Tables

Tables provide the most efficient way to display a large amount of data in a small amount of space. In proposals, tables tend to be used in two ways. First, they are used in Background sections to provide baseline data that creates a numerical snapshot of the past or the present situation. Figure 12.3, for example, might be used to show how bird populations have fluctuated in Story County, Iowa, in the past few years. Second, tables are used to present budgets, breaking down costs into cells that can be easily referenced in the written text.

   Labeling tables is handled a bit differently than labeling other kinds of graphics. The number and title should always appear above the table. Also, in most documents, figures and tables will be numbered separately. One set of numbers is used for graphs and photos (e.g., Figure 1, Figure 2, and so on), and a separate set of numbers is used for tables (e.g., Table 1, Table 2, and so on).

   As shown in Figure 12.3, the left column in a table usually includes *row titles* that list the items being measured. Along the top row, the *column titles* usually list the qualities or dates of the items being measured. Beneath the table, if applicable, a citation should identify the source of the information.

If a customer, client, or funding source requests a specific citation style (like APA, IEEE, or MLA), you should follow that citation style's guidelines for formatting tables.

Figure 12.3: A Table for Showing Data

TABLE 2: Counts of Perching Birds at Wilson State Park, February

Birds Sighted from 6:00-9:00 a.m.	2021	2022	2023	2024
Robins	370	380	420	443
Purple Finches	67	69	44	55
Cardinals	84	92	100	113
Blue Jays	34	45	23	76
Goldfinches	27	33	25	37

Source: A. Lawler (2025). "Habitat Restoration Strengthens Perching Bird Population." *Central Iowa Ornithology Quarterly* 73, pp. 287-301.

Pie Charts

A pie chart can be used to demonstrate how a whole can be divided. The pie chart in Figure 12.4, for example, shows the leading causes of death in year 2021 in the U.S. The story told by the pie chart is that heart disease and cancer together caused about half of all deaths in the U.S. in 2021. Another interesting story, though, is that COVID-19 was responsible for more than 1 out of 8 deaths that year (about 12.5 percent), which made it the third highest cause of death.

Keep in mind that pie charts use a large amount of space in a document to display only a small amount of data. For instance, the pie chart in Figure 12.4 uses a significant amount of space to show only ten data points. If you decide to use a pie chart, make sure the story you are illustrating is worth that much space in your proposal.

Figure 12.4: Pie Chart That Illustrates Percentages

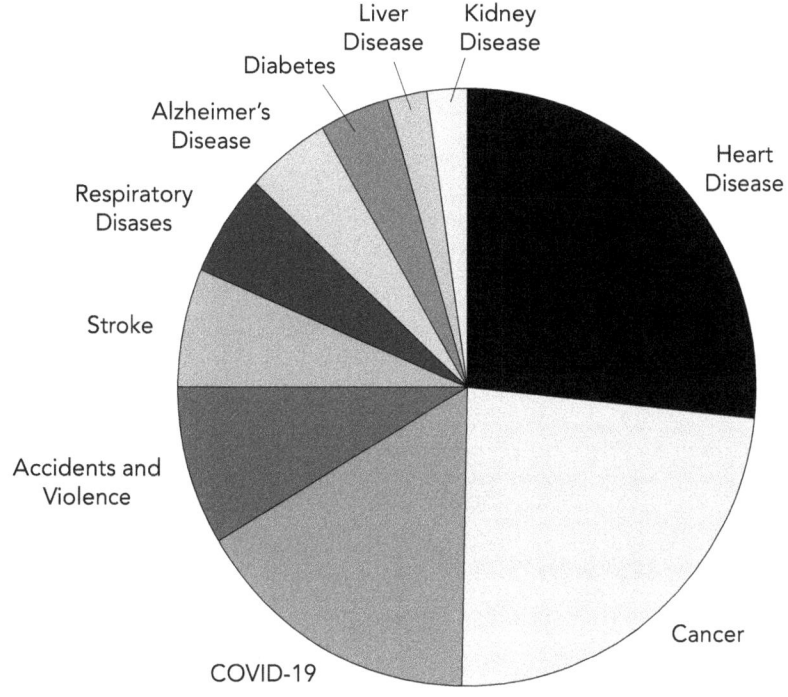

Source for Data: National Center for Health Statistics [NCHS]. (December, 2022). Mortality in the United States, 2021. NCHS Data Brief No. 456. Centers for Disease Control and Prevention.

Gantt Charts

Gantt charts are popular in proposals and grants, especially now that project-planning software can quickly generate these graphics. A Gantt chart, like the one in Figure 12.5, can be used to illustrate a timeline, showing when various phases of the project will begin and end.

There are two primary benefits to including a Gantt chart in a proposal or grant. First, the chart shows how various stages of a complex project will overlap and intersect. Second, it gives the readers an overall sense of how the project proceeds from start to finish. The Gantt chart in Figure 12.5 demonstrates both benefits.

Your readers will often expect a Gantt chart in business proposals, especially in technical fields. Increasingly, Gantt charts are also being used in grant proposals to show the steps or phrases of large projects. These charts are typically used to reinforce the explanation of the proposal's plan, offering a visual sense of how the project will progress.

Figure 12.5: A Gantt Chart

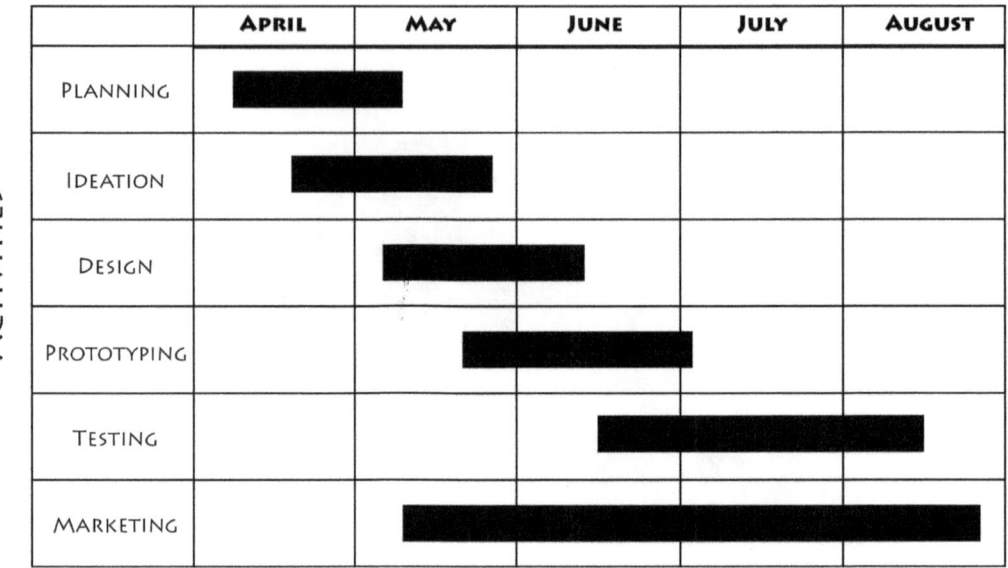

Organizational Charts ("Org Charts")

Organizational charts are used to illustrate the relationships among people in companies and organizations. Usually, these charts are placed in the Qualifications section of a proposal or an appendix, though occasionally they may be used within the Project Plan section to show who answers to whom. Essentially, organizational charts are designed to illustrate the chain of command for a project.

These charts can be helpful, especially when many people and teams will be involved in a project. However, as discussed in Chapter 7, organizational charts usually aren't worth the space in smaller proposals.

When deciding whether to include an organizational chart, ask yourself whether the chart serves a specific purpose. Does the organizational chart tell the readers something they need to know to make a decision? If your organizational chart does not have a clear purpose, move it to an appendix or don't include it in the proposal.

Turn to Chapter 7 about Qualifications sections to learn more about using organizational charts in proposals.

Photographs

Phone cameras, digital cameras, and scanners make inserting photographs into proposals and grants easier than ever. When used properly, photos can visually reinforce your points in the written text. But, much like the pictures from your last vacation, photographs don't always capture the essence of what you are trying to show. In some cases, photographs can leave your readers wondering what story you are trying to tell.

To use a photograph appropriately, make sure it tells a straightforward story. Meanwhile, resist the urge to use photos to merely decorate the proposal. If you decide to include images, the readers should be able to determine what story the photograph is trying to illustrate.

Stock photos can be purchased from suppliers on the Internet if you want something taken by a professional. Some stock photos are royalty-free, but most are not. If you use someone else's photo without permission, they can sue you for a substantial amount. So, make sure you contact the owner of a photograph before using it. Note: According to U.S. copyright laws, copyrighted photos can be used without permission for educational purposes. However, don't post your document on the Internet with these images unless you have permission.

AI applications can also create realistic "photographs" for you. The copyright laws are currently in flux, so it's not always clear who owns an AI-generated image. Keep in mind that AI image generators are trained on existing photos. If use an AI-generated image that is similar to one owned by someone else, they may be able to sue you.

Similarly, your readers may expect you to disclose any AI-generated images in your proposal or grant. Using AI-generated images without disclosing that they are synthetic can damage your credibility or that of your organization. In short, exercise caution and know your readers' expectations when using AI-generated images.

Drawings

Line drawings are often better than photos for illustrating machines, buildings, and designs. Photos usually include more detail than needed and don't reproduce well in some documents. A drawing can simplify the image to the basic features of your subject. It will also allow you to show close-ups of the parts in a machine, the details of a building, or the schematics of electronic components. Drawings can show *cutaways* of buildings or machines, illustrating features that are not visible from the outside.

Unless you are an artist, you should probably hire an illustrator to create drawings for your proposal. Before meeting with the illustrator, sketch (thumbnail) the drawing you have in mind. Then, when you meet with the illustrator, explain what story you want the drawing to tell or what point you want it to make. With this information, the illustrator will be better able to draw an illustration or diagram that suits your proposal's needs.

The drawback of using drawings is the time and expense required to create them. So, you should use them only when needed to make an important point in the proposal. If you're making a common point or just want to add a visual feature to the document, a drawing may not be worth the time and expense. In some cases, you may be able to annotate a photograph of the machine or building. Doing so can save you a lot of time and money.

Other Kinds of Graphics

Of course, graphics can be used to present information and data in countless ways. The graphics discussed in this chapter are commonly used in proposals and grants. You may also use radar plots, maps, flowcharts, blueprints, scatterplots, pictographs, logic trees, screenshots, and other graphics.

Each graphic discussed in this chapter can be altered to tell various stories. A bar chart alone can use a variety of formats, such as a horizontal bar chart, stacked bar chart, 100-percent bar chart, and deviation bar chart. If you need to create a specialized graph or chart, you can search for advice on the Internet or hire a graphic designer who specializes in making these kinds of graphics.

A Note about AI Images

Today's AI applications can quickly create impressive visual images that depict people, places, products, and services that aren't real. You can experiment with these applications, especially if you want to generate placeholder images for your proposal or grant. But don't use AI-generated images as evidence or support for your claims.

When using AI to create a graph, chart, or diagram, you should verify the AI's output to ensure accuracy. You should verify that the graphic accurately reflects the original data to ensure the AI application interpreted the data correctly.

Using AI-generated images without disclosing that they are synthetic is unethical and could damage your credibility. Instead, use real photos from your organization or purchase stock photos from a collection. Readers will appreciate the authenticity of real photography.

Chapter Summary and Looking Ahead

As you draft your proposal or grant, look for places where graphics might be used to reinforce the message in the written text. Each graphic can be roughed out while you are writing. That way you will have time to make the graphic yourself or hire someone to do it for you. If you wait until the proposal is fully drafted to create the graphics, you will run out of time. In today's competitive environment, the absence of graphics in a proposal or grant can often be the difference between a successful grant and one of the also-rans.

Keep in mind that our society is becoming increasingly visual. Graphics can be used to reinforce your main points, help your readers make comparisons among ideas, and illustrate trends. The adage that "a picture is worth a thousand words" won't allow you to replace a thousand words of your proposal with a photo, but graphics can clarify or drive home important points with your readers. They are powerful tools in proposals.

Try This Out!

1. Find a graph, chart, or table in a printed or online document. Does the graphic tell a simple story? Does it reinforce the written text, not replace it? Is it ethical? Is it labeled and placed properly? Then, find one or more graphics that do not follow these four guidelines. Are they still effective? How could you improve each graphic?

2. Find a document that includes minimal or no graphics. Looking through the document, can you identify where a graphic might have helped reinforce or clarify the written text? Sketch a few graphs, tables, or charts that might be useful in this document. Find a photo on the Internet that would have added life and color.

3. Use the data in Figure 12.3 to create two different kinds of graphs. Try making a bar chart and a line graph. How do these different graphs allow you to tell a different story with this data set? Which graph do you think is more effective?

4. Create a simple survey that asks questions about a local problem. Ask a few people to complete your survey (perhaps people in your class). Then, use the data you gathered to create a few graphs, tables, or charts.

5. Look closely at the photos on your favorite news source's website. Can you describe what stories the images were designed to tell, even if they were not accompanied by written text? Also, look for a photo that does not tell an obvious story. How could the author have used a better image to suit the needs of the document and its readers?

6. Find a professional document that appears to use AI-generated graphics. Do these graphics follow the four guidelines discussed in this chapter? Do you find these graphics persuasive? Would the intended readers find these graphics persuasive? Why or why not?

Case Study: The Carbon Neutral Campus Project— What Kinds of Visuals Would Help?

Tim jumped at the chance to design the graphics for the Carbon Neutral Campus proposal. Creating the graphics sounded more fun than doing the budget or editing sentences and paragraphs.

He and Calvin met to talk about the page designs of the grant proposal, giving Tim a better sense of the kinds of tables, graphs, and photographs he would need to find or create. He felt that pictures and drawings, especially, would make the text seem more realistic to the readers. Also, he thought graphs would be a good way to illustrate energy usage trends on campus.

The readers at the Tempest Foundation probably wouldn't be energy experts, so Tim guessed they would rely on the graphics to understand some of the more complex ideas in the proposal.

He began by taking photos on campus and then collecting others online. Taking photographs was not difficult. He grabbed his phone and hopped on his bike. He decided to take a few pictures of the campus and the Young Power Plant. These photographs would help the readers visualize what Durango University was like.

He also collected pictures of alternative energy sources from the Internet to show what kinds of renewable energy the campus might use.

He texted George, the engineer on the Carbon Neutral Campus grant-writing team, to ask his advice. George said Tim should look for photos and drawings on U.S. government websites. Copyright law would allow the team to use these government images as long as they were properly cited. Photos owned by the U.S. government are not copyrighted and can be used without permission.

Searching on government websites, Tim soon found several photos of alternative energy sources that he could use. He also came across a picture of a plug-in hybrid bus built by owned by National Car Rental. He wondered if that picture could appear in the grant proposal, so he texted George again.

"That's a little more difficult," replied George. "You will need to ask National permission to use that picture because they own it. But there's a good chance they'll give us permission to use it. Why don't you email their public relations department and ask? They probably wouldn't mind the free publicity."

Tim downloaded the picture of the plug-in hybrid bus and wrote an email to National asking permission to use it. He explained how the image would be used in the proposal and how it would be cited.

A few days later, he received an email message that gave him permission to use the photo at no cost. National only requested that the company receive credit for the photo in the proposal.

Tim began thinking about graphs and charts that could support the proposal's written text. On a sheet of paper, he wrote down the trends he wanted to illustrate:

- Show the increased energy usage on campus.
- Show the increased cost of energy on campus.
- Show broad support on campus for converting to renewable energy sources.

He realized the only place that would have figures on energy usage and costs was the Department of Physical Facilities, supervised by Anne Hinton, one of his team members on the Carbon Neutral Campus grant-writing team.

Tim texted her. She replied, "Oh, sure, we can get you that data. I'll send you our annual report. Then, if you want other reports with more data, I can send them to you."

Tim asked Anne how he might generate some numbers to show broad support for the Carbon Neutral Campus Project.

Anne said, "Maybe we can email an online survey to university administrators from the department heads up to the university president. The survey wouldn't be scientific, but we could use the data to indicate the amount of support for going carbon neutral with major decision makers."

"That sounds like a great idea," texted Tim.

Anne replied, "OK. You can email me your survey questions and some text to introduce the survey, and I'll have one of our staff members create a Qualtrics survey. When the survey is ready, I'll send an email asking the deans and chairs to complete the survey. Not everyone will take the time to do it, but we should get enough responses to generate useful data."

The survey Tim sent Anne included the four questions shown in Figure 12.6.

Figure 12.6: Tim's Unscientific Survey

Survey on Energy Usage on the Durango University Campus

The Department of Physical Facilities is conducting an opinion poll about energy usage on campus. The campus's increasing energy bills have been responsible for many of our budget constraints. Climate change is also an important concern. So, we are seeking alternatives to our present energy sources.

To gauge your impressions of the issue, we have created a simple survey that will provide the university with some preliminary feedback. Please answer the following brief questions. We don't need names or specific information, just your responses to the questions.

Thank you for your help. When you click 'Submit' the survey will be sent to us automatically. If you have any questions, please call me at (970) 555-1924 or e-mail me at anne.hinton@durangou.edu.

Dr. Anne Hinton
Vice President for Physical Facilities

Questions	Strongly Agree	Agree	Disagree	Strongly Disagree
The energy problems at Durango U. need to be addressed now, not later.	❏	❏	❏	❏
Our energy problems can be solved primarily through conservation of energy.	❏	❏	❏	❏
Durango U. should convert to renewable energy sources as soon as possible.	❏	❏	❏	❏
Durango U. has an obligation to take on a leadership role in the region regarding energy issues and issues of global warming.	❏	❏	❏	❏

A week later, Anne sent Tim the data from the survey. Over 55 percent of the administrators had responded. Tim tabulated the responses to the survey questions and entered them into his spreadsheet program. He decided to display the responses as a vertical bar chart, which illustrated the answers to each response (Figure 12.7).

Tim then used the annual report Anne sent him to do more research on Durango's three primary energy sources: natural gas, electricity, and petroleum. Based on data from the annual report, Tim used draw a line graph showing the significant rise in energy costs on campus. Now he understood why President Wilson was so worried about rising energy costs on campus. They weren't just rising: They were skyrocketing!

Figure 12.7: Tim's Bar Chart Based on Survey Results

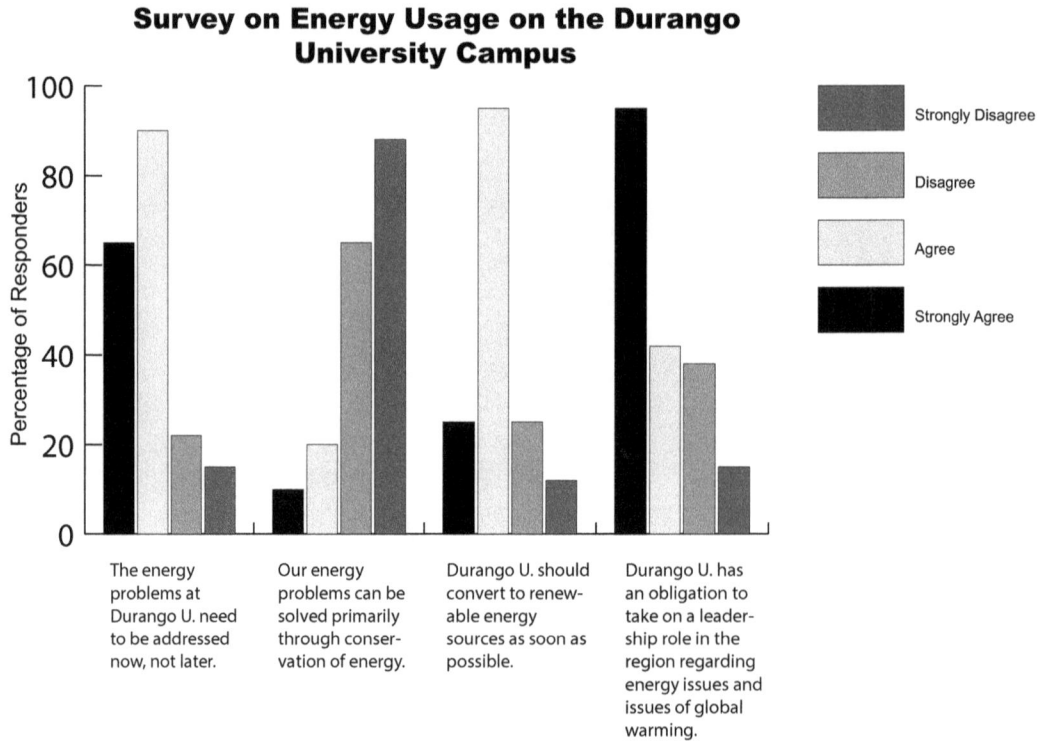

He needed a couple of hours to finish making the graphs, and he collected all the photos in one folder on the team's document-sharing site. Then, he texted Calvin, who was designing the document, to show him what kinds of graphics he wanted to include in the proposal.

Tim realized that the time collecting photos and making graphs was time well spent. His visuals would make the grant proposal seem much more realistic and professional.

If you would like to see how the graphics in the Carbon Neutral Campus proposal turned out, please go to Appendix A.

13 PUTTING YOUR PROPOSAL TOGETHER

Seeing the Proposal as a Whole Document

In this book, you've learned how to draft and design proposals and grants using rhetorical techniques. As you near the end of the proposal-writing process, it's time to begin seeing your proposal as a whole document. You should ask yourself whether your proposal achieves its purpose and meets the readers' needs. These are tough questions, especially when you have a nearly finished draft of the proposal on your screen. This last phase, when you are assembling the final version, is when you will take your proposal from "good enough" to "really strong."

In this chapter, you will learn how to assemble your proposal as a package. The proposal is the core of that package, but you will also include additional features, such as a letter or memo of transmittal, cover page, table of contents, executive summary, and appendices. These features are called the *front matter* and *back matter*. As you assemble the package, you should save time to revise your proposal at least one more time. While revising and editing, you will add the extra refinement that will persuade your customers, clients, or funding sources to say yes to your proposal.

Preparing the Front Matter

Front matter consists of the materials that appear before the introduction of the main proposal. The front matter is intended to set a context or framework for the proposal. Items in the front matter include some or all the following items:

- Letter or memo of transmittal
- Cover page
- Executive summary
- Table of contents

These materials are not mandatory unless requested by your customer, client, or funding source. Nevertheless, they are often helpful accessories that help the readers work more efficiently through your ideas. Including these front matter items also shows attention to detail and professionalism, which your readers will appreciate.

Grant proposals may also require you to fill out forms that provide information about your organization, its nonprofit status, and other information. These forms are usually self-explanatory, but the point of contact (POC) at a funding source can answer any questions about filling them out.

Letters or Memos of Transmittal

The purpose of a letter or memo of transmittal is to introduce the readers to the proposal or grant. Even though a proposal may not need a letter or memo of transmittal, you should include one for a couple of reasons. First, letters and memos are personal forms of correspondence, so they will add a personal touch to the proposal package. A letter or memo of transmittal allows you to introduce yourself and shake hands with the readers before they start reviewing your proposal. Transmittal letters and memos tend to be more personal in tone than proposals, setting the readers at ease before they start considering your ideas.

A second reason why letters and memos of transmittal are important is their ability to steer your proposal to the right person. In large companies and organizations, documents sometimes end up on the wrong desk, in the wrong inbox, and ultimately in the trash can or recycle bin. By identifying your proposal's readers and purpose up front, your letter or memo of transmittal can steer our proposal to the right place. And, if the wrong person happens to receive your proposal, the letter or memo of transmittal will help them send or forward it to the person who should receive it.

For grant proposals, funding sources often do not ask for a letter of transmittal, but this kind of letter can be a nice addition to the package. Readers at private foundations will often warm up to the personal touch of a transmittal letter. If they don't want the letter, they can remove it before the package is reviewed.

In many cases, your transmittal letter may take the form of an email with your full proposal or grant attached. Even if your transmittal letter is an email rather than a standalone document, you should still treat it as a personal correspondence that introduces the readers to the proposal and your company or organization.

Writing a Letter or Memo of Transmittal

Letters and memos of transmittal should have an introduction, body, and conclusion (Figure 13.1):

> **Introduction**—The letter or memo of transmittal's introduction should provide background information on your organization and state the purpose of the proposal. Specifically, you should identify the request for proposals (**RFP**) to which the proposal is responding. If the RFP includes a reference number, you should include it here (and in the subject line if you're submitting through email). This kind of opening will help the readers remember exactly why the proposal is being sent to them. You can also tell the readers the purpose of the proposal. You can even use the same purpose statement that appears in the proposal's introduction, though most writers will paraphrase that sentence to avoid the feeling of repetition. An introduction to a letter or memo of transmittal should run, at most, a few sentences.
>
> **Body**—The body of a letter or memo of transmittal should highlight and summarize the proposal's main points. You might describe your plan in miniature if you think it will

grab the readers' interest. You might also mention some of the key benefits of your plan or briefly describe the qualifications of you, your team, or your company. Try not to run on more than a page; a letter or memo of transmittal should be concise, saving the details for the proposal itself. As a rule of thumb, the body of this letter or memo should run about two or three short paragraphs.

Conclusion—The conclusion should thank the readers for their time and tell them who to contact if they have questions or need further information. You might also tell the readers that you will contact them by a specific date to follow up on the proposal. The conclusion should be concise, perhaps two or three sentences.

Figure 13.1: Format for a Letter of Transmittal

Company or Organization Letterhead

Address, Phone Number, Web Address

Date Proposal was Sent

Point of Contact Name (if available)
Customer's or Funder's Name
Street Address
City, State, Zip Code

State the purpose of the transmittal letter. Identify the request for proposals (RFP) by title and number. State the main point of the proposal.

Your company's or organization's background and qualifications. Highlight your company's or organization's experience and a quality that sets it apart from competitors.

A one or two sentence overview of the proposal. Three to five reasons why the proposal is competitive for the project or funding.

Thank the readers for their time and attention. Provide a name of someone they can contact and include that person's contact information.

Sincerely,

Handwritten Signature

Printed Name
Title of Sender

Using Style in the Letter or Memo of Transmittal

Stylistically, letters or memos of transmittal should be simple and personal. The tone should be upbeat and friendly yet respectful. A judicious use of *you* will make a personal connection between the readers and yourself.

Meanwhile, avoid using business-letter cliches like "enclosed please find," "pursuant to our agreement," or "as per your request." This kind of hackneyed business language will only make your letter or memo sound distant and aloof, which is not the tone to use if you want the readers to trust you.

Instead of using cliches, simply write the letter as though you are talking to another person. After all, you *are* talking to another human being, not a corporation. In a face-to-face conversation, you would never use these tired cliches and strange phrases, so avoid using them in a letter or memo.

Cover Page

The cover page of a proposal should identify the proposal's topic and set a particular tone. The cover page typically includes the following items:

- Title of the proposal
- Name of the client's company or the funding source
- Name and logo of the company or organization submitting the proposal
- Date on which the proposal was submitted

Cover pages can be simple or elaborate. Like the rest of the proposal, the cover page's design should fit the project's character and the readers' needs. For example, the cover page on the left in Figure 13.2 is intended to set a conservative tone for a proposal. Its balanced page offers a feeling of security. The cover page on the right is more progressive, using a unique page layout. An artificial intelligence (AI) application can help you create decorative drawings or a logo for the cover page if you want to set a particular tone.

Is a cover page required? No, not unless the customer, client, or funding source requested one. However, a cover page is an easy way to set a professional tone for the proposal. Like the cover of a book, cover pages provide a distinct starting point at which the readers will begin assessing the merits of your ideas.

Executive Summary

In our fast-paced culture, executive summaries are expected in proposals and grants, especially larger ones. An executive summary boils the proposal down into a synopsis that can be read in minutes. In one to three pages, the executive summary condenses the background, the project plan, the qualifications, and the costs and benefits. It provides a snapshot of the proposal so your readers can quickly determine its major points and claims.

Figure 13.2: Two Different Designs for Cover Pages

Executive summaries tend to follow the organization of the proposal itself. The first paragraph in the summary identifies the proposal's topic, purpose, and main point. The body paragraphs summarize the background, project plan, qualifications, and costs and benefits.

Fortunately, AI applications are really good at writing summaries of documents. You can use AI to help you write an executive summary of your proposal or grant. Before copying your proposal into a third-party AI application or uploading it as a file attached to your prompt, you should remove proprietary or personal information. (Caution: Assume that any information you copy into a third-party AI application will be harvested by the company providing it.) Then, copy your proposal into the AI application or upload the file as part of your prompt. Next, ask it to create an executive summary of a general length (e.g., "around 500 words").

You can then paste this AI-generated executive summary into your proposal or grant. Revisions will almost surely be needed. You can then re-insert the proprietary or personal information you removed earlier. Using AI will save you a considerable amount of time. You might even find that the AI-generated summary offers a more straightforward explanation of your proposal than the current draft. In these cases, you can use the language from the AI-generated summary to clarify some of the phrasing in your proposal.

Don't underestimate the importance of the executive summary. The executive summary may be the only version of your proposal that some readers actually read. Decision makers at most companies and funding sources are typically short on time, so they often rely on executive summaries to help cut down the pile of competing proposals to a few finalists they will read in depth. Also, readers use the executive summary at meetings to help them quickly refresh their understanding of the proposal's main points.

In many cases, the decision makers may end up reading the executive summary more closely and frequently than the proposal itself, so it needs to be well written. If you use AI to help you summarize your proposal, review its output closely to ensure its content and style are accurate and appropriate.

Table of Contents

A table of contents is a standard feature in most proposals, especially ones that run more than ten pages. The table of contents provides the readers with an outline of the proposal's major sections, helping them determine the most efficient way to read the text. This overview also forecasts the structure of the proposal, providing the readers with a mental framework for understanding how the proposal is organized. Each title in the table of contents should be the same as the headings for the major sections.

Most word processing applications, like MS Word or Google Docs, can generate a table of contents when the document is completed.

Preparing the Back Matter

Back matter includes additional materials that support the information in the body of the proposal. Often called the appendixes, the back matter in a proposal becomes a reference tool for the readers. Back matter can include any or all the following items related to the proposal:

- Itemized budget and budget rationale
- Resumes of management and key personnel
- Analytical reports or white papers
- News or magazine articles
- Prior proposals
- Formulas and calculations
- Glossary of terms
- Bibliography
- Personal or corporate references

Each appendix should be labeled with a number or letter so it can be referred to in the body of the proposal. For example, in the body of the proposal, the readers might see the note, "The geothermal formulas and calculations used to determine these figures can be found in Appendix D." If the readers want to check those figures, they will know that they can then turn to that appendix in the back of the proposal.

In each appendix, an opening paragraph should introduce the contents that follow. For instance, if you decide to include magazine articles you collected on the topic, the appendix's opening paragraph might explain that you have included the articles to show the issue's importance to the public or industry. You might also briefly discuss how these articles reinforce your argument in the proposal's body. Even a glossary of terms should include some kind of short opening paragraph to help the readers understand its purpose.

The most important item in the back matter, a budget rationale, was discussed in Chapter 9 on budgeting. Let's look at some other kinds of back matter.

Resumes of Management and Key Personnel

It is becoming increasingly common to include the resumes of managers, principal investigators, and key personnel in an appendix to the proposal. Though your Qualifications section may have included biographies of the executives and researchers involved with the project, the readers may want to know more about their backgrounds. One- to two-page resumes for each executive or researcher will help the readers gather this additional information.

A typical resume in the appendix will include the following information about the managers and key employees on the project:

- Employment history
- Education
- Publications
- Special training or skills
- Awards
- Memberships in professional groups
- Internet addresses or hyperlinks for LinkedIn sites or Open Researcher and Contributor Identifiers (ORCID) records

An appendix with resumes should also include an opening that states the purpose of the appendix (i.e., to provide resumes of managers and key personnel and list the people whose resumes follow).

Glossary of Terms and Symbols

As much as we try to avoid it, proposals and grants often include jargon words and symbols unfamiliar to the readers. In these cases, an appendix that contains an alphabetical listing of terms and symbols with definitions can help the readers work through the complex terminology of the proposal. Glossaries are especially helpful for readers who are not familiar with the technical aspects of the project.

Each item in a glossary should be written as a sentence definition. A sentence definition has three parts: (1) the name of the item, (2) the class to which the item belongs, and (3) the features that distinguish the item from its class. For example, here are a few examples of definitions that might appear in a glossary:

Cathode ray tube: a vacuum tube in which electrons are projected onto a fluorescent screen.

Cyclotron: a particle accelerator in which charged particles (e.g., protons) are propelled in a circular motion by a magnetic field.

Direct labor: the group of workers who are directly active in the production of goods and services.

Σ (sigma): the total number of electrons measured after one minute of testing.

Each of these definitions starts with the name of the item being defined. Then, it identifies the class to which the item belongs (e.g., particle accelerator, vacuum tube, body of workers, total amount of electrons). The remainder of the definition explains how the item can be distinguished from its class.

Bibliography or References List

A bibliography or references list should include a list of print or electronic sources, interviews, and other outside materials cited in the proposal. It should also list the sources of any quotations, graphics, data, or ideas the proposal uses from another document. Also, if you conducted interviews, place the dates and times of these interviews in the bibliography.

The format of a bibliography or references list should follow guidelines found in style manuals, such as those offered by the American Psychological Association (APA), the Institute of Electrical and Electronics Engineers (IEEE), or the Modern Language Association (MLA). APA style is widely used in scientific and technical proposals. IEEE is used in many technical documents. MLA style tends to be used in nontechnical proposals. Many reference books and the Purdue Online Writing Lab (OWL) include examples of these styles. However, it is always a good practice to consult the style guides themselves. Occasionally, corporations and government bodies will develop their own style guides, which provide specific rules for formatting and citations.

For most proposals, the citation style chosen is not all that important (unless the customer, client, or funding source requests a specific citation style). Consistent formatting of your citations, however, *is* important. Each item in your bibliography or references list should follow a predictable format. Though formatting bibliographic entries correctly may seem a trivial concern, dedicating time and attention to citation practice indicates to your readers that you and your organization pay attention to details. Research management applications (such as Zotero or BibDesk) can also help you prepare a correct and consistent bibliography or references list.

Formulas and Calculations

In the body of a proposal or grant, formulas and calculations can be a momentum killer for your readers. And yet, these numerical tools are critical if your readers want to assess the soundness of your proposal. In a research grant, formulas and calculations are important but can overwhelm nontechnical readers.

You may consider moving some or all your formulas and calculations to an appendix, especially when addressing an audience that includes nontechnical readers. In the appendix, each calculation or formula should be labeled properly in a way that explains how it is used. In some cases, an explanatory narrative should be included that walks the readers through any calculations or derivations of formulas. As much as possible, try to avoid dumping the readers into a maze of symbols and numbers. After all, if the customer, clients, or their experts cannot follow your calculations or derivations, they may not trust the results and conclusions you draw from them.

Related Reports, Prior Proposals, and FYI Information

You might include copies of documents discussed in the proposal. For example, a scientist may include an article or white paper that supports an important argument in the grant proposal. A business proposal writer may include a recent magazine article that praises their company. Also, you may want to include prior proposals related to the topic.

This appendix can include any additional documents as long as the readers are not overburdened with unnecessary information. For each document you add, you should also include a small paragraph that identifies the background for the document. Tell the readers why the document is important, when it was written, who it was written for, and where it was used or appeared. This explanatory information is important because you should not expect the readers to figure out for themselves why you decided to include additional materials in the appendix.

Revising the Proposal or Grant

In Chapter 1, you learned that writing a proposal involves a process. Even the best proposal writers and grant writers cannot sit down and crank out a proposal in one try—though you will certainly hear myths about the person who wrote a winning proposal in one day. Myths aside, proposals usually require many days and several drafts before they are complete. To succeed, you should devote significant time to revising and editing your documents.

Most writers revise their work as they finish each section of a proposal–always a good idea—but you should also rethink and revise the proposal as a whole text. Why? Often, drafting a proposal causes you to think deeply about your topic. Your original ideas will have evolved or changed about the current situation, the plan, and the readers' needs. While revising, you can revise all the proposal's major sections, so they tell a consistent story.

The Rhetorical Situation

Start the revision process by reviewing your notes about the rhetorical situation. Ask yourself the following questions, looking at each section in the proposal separately. Start by reviewing the topic. Ask yourself the following questions:

- Is the topic of the proposal still the same?
- Does any content in the proposal drift outside the boundaries of the topic?
- Is the topic at the beginning of the proposal the same as the topic at the end?

Next, review the purpose. Ask yourself the following questions:

- Does the proposal achieve the stated purpose?
- Did the purpose of the proposal evolve as the proposal was drafted?
- Do you need to refine or broaden the purpose statement to suit a deeper understanding of the situation?

Now consider your readers by answering the following questions:

- Does the proposal address the primary readers' motives, values, attitudes, and emotions?
- Will the secondary and gatekeeper readers be satisfied with the proposal?
- Does the proposal contain any information that hostile tertiary readers could use to damage you, your team, or your company or organization?

Finally, use the following questions to reexamine the context:

- Is the proposal appropriate for the physical, economic, political, and ethical situations in which it might be used?
- What changes to the style, format, or design would make it more usable?
- Are any parts of our proposal politically or ethically problematic?

You need to be honest with yourself as you revise and edit the final draft. Writers sometimes ignore problems with the proposal now that the finish line is in sight. Coworkers may be hesitant to bring up mistakes or contradictions because they don't want to upset others. But you should keep reminding yourself that the readers will not ignore these problems. The extra time and effort you spend rethinking the rhetorical situation will help you make the final adjustments needed to help you win the contract or receive the funding you need.

Rethinking the Problem or Opportunity

After reconsidering the rhetorical situation, look closely at the notes where you identified the problem or opportunity the proposal or grant was written to address. Often, your original understanding of the problem or opportunity will evolve during the drafting and revision phases, especially in larger proposals. As a result, the proposal is now handling a much more complex issue than was first defined. If so, perhaps you need to scale back the proposal to address a more targeted problem or opportunity. Or maybe you need to rewrite parts of the proposal to suit that larger problem.

Ask yourself again, "What changed?" as you reconsider the problem or opportunity. As mentioned in Chapter 1, proposals and grants are tools for managing change, so while revising the proposal, you should ask yourself whether your proposal addresses the elements of change that are driving the current situation. Are you shaping change to the advantage of your company or organization? Are there any elements of change that you are not addressing in your proposal?

Then, think about the type of proposal or grant you were asked to write. Were you supposed to write a research proposal, planning proposal, implementation proposal, estimate proposal, or a combination of these kinds of proposals? Often, during the proposal-writing process, the project concept experiences some mission creep. For example, a research proposal will tend to creep ahead and start offering a strategic plan, or a planning proposal will begin describing implementation, providing unnecessary details like steps, names of personnel, and schedules that typically would be found in an implementation proposal.

To avoid mission creep, do your best to give your customers or clients exactly what they asked for—nothing more and nothing less. Customers and clients usually want to do a project one step at a time. If they asked for an assessment of their facilities' production capacity, they don't want you to describe how you would also rebuild their manufacturing facility. If they asked for a cost estimate for recycling post-production materials, they don't want a planning proposal that shows how you would restructure their assembly line to reduce waste.

Another reason to give the readers exactly what they want is that your ideas shouldn't be given away for free. If you provide them with a detailed implementation plan that goes beyond the strategic plan they requested, you just gave them a blueprint for solving their next problem—without paying you!

The Final Edit

As you near the first final version of the proposal, pay attention to the content, organization, style, and design.

Content

- Is the proposal's content complete?
- Is there any information missing that would help your proposal persuade the readers?
- Are there any digressions in the proposal where you have included details that go beyond need-to-know information?

Organization

- Does the proposal follow a logical order or the order specified by the customer, client, or funding source?
- Does the proposal tell a story, leading the readers from the current situation, through a plan, to a beneficial conclusion?

- Does the introduction set a compelling context for the body of the proposal by highlighting the topic, purpose, and main point while offering helpful background information, stressing the importance of the topic, and forecasting the body of the proposal?
- Do the opening paragraphs of each section identify the purpose and point of their respective parts of the proposal?

Style

- Does the proposal reflect an appropriate tone/persona for the readers and topic?
- Is the style plain where the proposal instructs the readers and persuasive where you are trying to influence them?
- Can you use persuasive stylistic devices like similes, analogies, or metaphors to make the text more visual or powerful?

Design

- Does the proposal follow an appropriate, consistent page design?
- Are the pages balanced?
- Does the proposal use alignment, grouping, and consistency appropriately?
- Have you used bulleted or numbered lists where appropriate?
- Does data appear in tables or charts?
- Are the graphics properly placed and labeled?
- Would more graphics help illustrate difficult points?

As the deadline looms and you grow tired of working on your proposal, it's tempting to cut the revision time to a minimum. You might even convince yourself that the readers don't care whether the proposal is polished. Actually, readers care very much about the final form of the proposal. It's the only version they will ever read, and errors make it seem unfinished.

Often, the difference between the successful proposal and the runners-up is the added time the winners put into revising the final document. Sayings such as "the devil is in the details" and "quality is found at the edges" are true of proposals. Always set aside a significant block of time to revise the proposal. The additional time spent on revising can often be the difference between winning and losing the contract or funding.

Looking Ahead

A proposal or grant is the beginning of a relationship. Essentially, your readers are interviewing your company or organization to determine whether a basis for a positive, constructive partnership exists. Your proposal is the face you are presenting to the customer, client, or funding source. If they feel comfortable with your proposal, they will feel comfortable with your company or organization. If something feels wrong about your proposal, they will choose another proposal that "just feels right."

In this chapter, we went over the endgame of the proposal-writing process. Once you finish with the proposal, you should be able to see the document as a whole. It should be complete, organized, easy to read, and well-designed. If you feel comfortable with the proposal in its final form, with no regrets, chances are that your readers will feel comfortable with it, too.

Try This Out!

1. Look over the two proposals included in Appendix A. How do they meet the purpose and needs of their readers? Are there any further revisions you might make to these proposals? How could they be improved?

2. Find a proposal or grant on the Internet or at your workplace. Write an analysis of the proposal in which you discuss the document's content, organization, style, and design. In your analysis, highlight examples of the proposal's strengths, and then make some suggestions for improving the parts of the proposal that could be stronger.

3. Write a letter or memo of transmittal for a real or practice business proposal or a grant proposal of your own. What information do you believe belongs in this letter or memo? What information should you save for the proposal itself? How can you write the letter with a positive, personal tone that puts the readers at ease?

4. Using a proposal from the Internet or your workplace, write a one-page executive summary describing the proposal. Your summary should cover all the major sections of the proposal. What did you decide to include? What did you leave out? How did you decide what to keep and what to leave out?

5. Use an AI application to summarize a short document you know well. Is the summary accurate? Did the AI include only need-to-know information? Did it add any information? Where did the AI fall short? Using your own proposal, work through the revision process described in this chapter. Revise your work by reconsidering the proposal's topic, purpose, readers, context, and objectives. Then, check whether your understanding of the problem or opportunity has evolved as you wrote the proposal. Finally, edit the proposal by paying special attention to its content, organization, style, and design.

Case Study: The Carbon Neutral Campus Project—Are We Really Finished?

The Carbon Neutral Campus team decided to meet one last time to finish the proposal. As they put the sections of the proposal together, they realized that some of their ideas about the subject, purpose, and readers had evolved. They had developed a sharper understanding of the problem they were trying to solve and its potential solutions. George's budget and Tim's graphics added some new information that needed to be addressed throughout the grant proposal. Overall, though, they had come a long way since their first meeting.

They began revising the grant proposal by reviewing their notes about the rhetorical situation. They edited the proposal, marking places where they may have strayed from the subject or purpose. They also thought carefully about whether they were addressing the concerns and needs of the reviewers at the Tempest Foundation.

"It's interesting," said Calvin. "We began by talking about all the environmental issues on campus, including recycling and water usage. But, as we worked on the proposal, we narrowed our subject down to energy issues."

Anne said, "That's fine. We were thinking a little too broadly at the start, which would have made our grant sound too complex to the reviewers."

George laughed, "This one is complex enough already!"

"We do seem to be dreaming big, aren't we?" asked Karen.

"Why not?" asked Anne. "If we were thinking small, we wouldn't need a grant. Anyway, a project like this one should appeal to the Tempest Foundation and other funding sources."

They worked on editing and proofreading the final version. With all the parts in one place, it was easier to see how to fix problems with the proposal's organization. They also revised or crossed out sentences that did not include need-to-know information.

Calvin made small changes in his page designs to help the written text work better with Tim's graphics. Calvin also edited for consistency, making sure the headings, rules, and lists were all handled in the same way.

As Calvin made the few last adjustments, George asked, "Are we finished?"

Tim said, "It looks done to me."

"Of course, I will be running a copy past President Wilson and some of our other gatekeeper readers, but it looks good for now," said Anne. "I'm sure they will have a few suggestions for changes."

"I'm glad we did this," said Karen. "I learned a great amount about energy, Durango University, and issues involved in the climate crisis. We took a big step in the right direction by writing this grant."

George agreed, "I hope this grant proposal is just the beginning of something special."

"Yeah, let's hope it is," said Calvin. "I really enjoyed working with all of you, and I look forward to working together again soon, especially if we get the funding!"

A final version of the Carbon Neutral Campus Project grant proposal is in Appendix A of this book.

APPENDIX A: CARBON NEUTRAL CAMPUS PROPOSAL

The Carbon Neutral Campus Project at Durango University:

A Grant Request to the Tempest Foundation

The climate crisis can seem overwhelming. A 2023 report from the Intergovernmental Panel on Climate Change (IPCC) found that human-related greenhouse gas emissions have continually risen over the past decade and that a global temperature increase of 1.5 C is all but inevitable. The IPCC indicates that the global climate crisis will likely have the following dire effects:

- Storms will increase in severity, posing a serious risk to life and property.
- Sea levels will rise between 50-100 cm compared to the 1900 baseline, causing significant flooding of lowland areas.
- Heat waves and droughts will be more frequent and severe in the United States.
- Many parts of the world will experience much heavier rainfall.
- Oceans will increasingly become deoxygenated and acidified, leading to a loss of oceanic resources.
- Many species of animals will go extinct due to loss of habitat.

The 2023 report from the IPCC notes that the evidence for the human-caused climate crisis is "unequivocal" and that "widespread and rapid changes in the atmosphere, ocean, cryosphere, and biosphere have [already] occurred." It warns that unless humans dramatically cut emissions of greenhouse gases, the impact of the climate crisis on this planet will grow to catastrophic proportions. These kinds of dire predictions might lead some people to feel hopeless. They may conclude that we cannot do anything locally about this complex global problem. The problem is too big.

At Durango University, we believe we can do something. Durango University has already made significant strides toward conserving energy. Still, we would like to fully convert our campus to sustainable non-carbon energy sources, which, as the IPCC (2023) notes, are more affordable than ever. So, with this proposal, we are turning to the Tempest Foundation for a grant to help us develop a Carbon Neutral Campus Strategic Plan. This strategic plan will guide the conversion of our campus to renewable and sustainable energy sources. With a long-term strategic plan in place, we believe Durango University's campus could become carbon neutral by 2035. Meanwhile, as we work toward converting our campus to renewable energy sources, we could help other universities worldwide follow our lead.

In this grant proposal, we will discuss Durango University's current energy usage and describe how we would use an urban planning charrette to develop and write the Carbon Neutral Campus Strategic Plan. We will discuss our qualifications for this project and its costs and benefits. The funding provided by the Tempest Foundation would allow us to develop the Carbon Neutral Campus Strategic Plan that could serve as a model for other universities. With this plan in place, we could then devote our own resources toward making the Carbon Neutral Campus Strategic Plan a reality.

Energy Issues at Durango University

Going green is not a new movement on our campus. Durango University began its Green Campus Program in 2001 with a grant from its own Office of the President. The Green Campus Program has already generated many benefits:

- increased conservation of university-owned lands
- heightened environmental awareness on campus
- increased recycling
- purchase of renewable energy certificates to offset 30% of the campus's electricity consumption
- use of compact fluorescent lightbulbs and Energy Star appliances

The Green Campus Program brought our community together and fostered a common environmental mission, but the long-term energy challenges we face at Durango University go beyond the scope of the Green Campus Program, which is now over two decades old.

At this point, the campus infrastructure itself is our core problem. The campus energy system was built in 1914 with the available energy technologies, specifically coal and oil. Without a broader vision in mind, our campus will continue to be bound to this carbon-based energy infrastructure, making us contributors to the climate crisis and vulnerable to increasing energy costs.

The Legacy of Old Betsy

Our campus's reliance on natural gas for steam heat is the most difficult problem to solve. In 1914, when the university was founded, a coal-fired plant nicknamed "Old Betsy" supplied steam heat and electricity from the eastern edge of campus where the Student Union is today (Figure 1). Old Betsy's coal-fired boilers kept the campus warm by pushing steam through underground tunnels running to the buildings. The plant also used steam power to generate the small amount of electricity required by the campus (Philips, 23).

Of course, the campus has changed significantly since then, but the technologies used to heat and power the campus are basically the same. The campus now has twenty-four buildings, which are spread over a campus that covers eighty-six acres. Old Betsy was replaced in 1931 by a larger coal-fired plant, and successively larger coal-fired plants were completed in 1954 and 1983. Each new or renovated power plant expanded the outdated infrastructure initially designed for Old Betsy (Philips, 63).

ᴊre 1. Old Betsy

So, the legacy of Old Betsy lives on today. Presently, the campus is heated by the Young Power Plant, a coal-fired plant built in 1983 and renovated in 2010 (Figure 2). It burns natural gas to make steam and then pushes it through tunnels under the campus. The campus's electricity demands outstripped the generating capacity of the Young Power Plant in 1995, mainly because of new electricity-using devices on campus like televisions and computers. Today, the campus draws 73% of its electricity from the Four Corners Power Plant, a massive coal-fired plant located west of Farmington, New Mexico. A small amount of our electricity (about 12%) is drawn from other regional power plants, which burn natural gas.

After thirty-five years of service, the Young Power Plant is due for replacement or a complete overhaul (Philips, 8). Its boilers are already expensive to run and maintain compared to other sources of electricity and heat (Games, 12). Steam is an inefficient way to generate electricity and heat a campus of our size. Moreover, burning natural gas adds carbon dioxide and other greenhouse gases into the atmosphere, contributing to the climate crisis.

Oil, Cars, and Campus

Our next most significant problem concerns the number of cars and trucks traveling to and around the campus daily. The campus was originally designed to be pedestrian friendly. By the late 1930s, however, automobiles were a common way to commute to and around campus. As a result, expansion plans for the campus began to centralize the automobile. Public transportation eventually disappeared. The campus trolley called the Dinky stopped operating in 1935, and the campus bus service was discontinued in 1978.

Plans for expanding the campus have routinely called for more parking lots and parking garages to accommodate the increasing number of cars. Streets have been widened, and major thoroughfares have been built to accommodate the greater traffic flow. These changes have allowed university employees and students to commute from even further away, thus causing a cycle of more parking garages, even wider streets, and

Today, the campus is over-reliant on automobiles and trucks for its transportation needs. The persistent complaints about a shortage of parking conceal the real problem: few alternatives to cars are available for commuting to campus. In the past, the university encouraged people to walk to campus or ride their bikes, but the campus design made driving a car more convenient and safer.

In the United States, transportation has been shown to be the single largest source of greenhouse emissions causing the global climate crisis (Bailey & Grossman, 2023). Electric vehicles (EVs), though carbon neutral to operate, can be prohibitively expensive for most of our students and staff members. In our region, EVs would be charged with electricity primarily from coal-fired plants. Our campus also lacks charging stations that would make EVs feasible, even if the costs were a nonissue. And the lack of public transportation requires even environmentally conscious people to rely on automobiles, which are often gasoline powered.

Effects of Inaction

Doing nothing is not an option for Durango University. The Young Power Plant will need to be completely overhauled or replaced within the next decade. The burning of natural gas, coal, and gasoline will only continue adding to the cascading problems of the climate crisis. Meanwhile, the accelerating energy costs are driving up expenses all around campus. If we do nothing, more of the university's budget will be used to pay for heating, electricity, and gasoline. These additional costs will be passed along to students, or they will lead to cuts in salaries, services, and staff. We also recognize that our reputation is at stake as other universities lead in using renewable and sustainable energy sources. Students are increasingly reluctant to attend universities that rely on "dirty" energy sources.

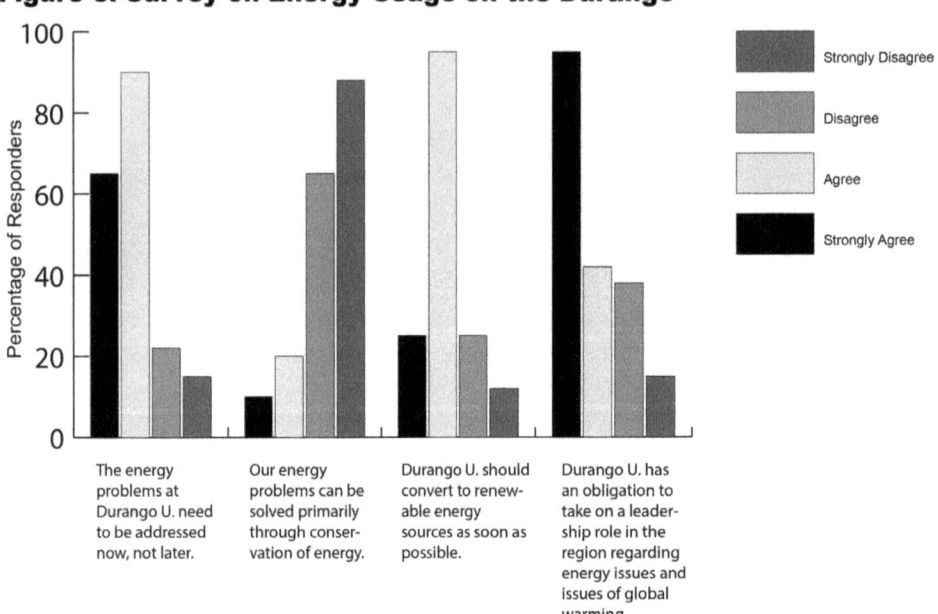

Figure 3. Survey on Energy Usage on the Durango

Durango University's Green Campus Program has raised our university's awareness of environmental issues on campus, but we are ready to make the kinds of large-scale infrastructure changes needed for a sustainable future. In a recent poll of our deans and department chairs, the support for this change was overwhelmingly positive (Figure 3). We believe this kind of support opens the door for achieving carbon neutrality.

Appendix A: Carbon Neutral Campus Proposal

Converting a college campus to renewable energy will take careful planning and time. Therefore, our primary goal is to develop a comprehensive Carbon Neutral Campus Strategic Plan to transform our campus into a net-zero carbon emission campus. To develop this plan, we will host an energy-focused charrette inviting the whole community into the planning process.

A charrette is a weekend retreat facilitated by professional urban planners that puts citizens and stakeholders into design teams (Figure 4). These design teams develop separate plans, drawing from their collective wisdom and community knowledge. Then, the urban planners use the ideas generated by these teams to create professional designs and present them back to the community. The advantage of a charrette is that it works from the grassroots up, using the community's knowledge, desires, and experiences, while encouraging all stakeholders to participate and buy into the project.

Our objectives in the Carbon Neutral Campus Charette would be the following:

- Develop a comprehensive strategic plan that would guide Durango University's efforts toward carbon neutrality while shaping future decisions about renovation and construction.
- Create and foster a community dialogue about renewable energy beyond the campus.
- Produce a new model of campus planning that shows how other campuses can use charrettes to work toward converting to renewable energy sources.

Step One: *Create the Carbon Neutral Campus Steering Committee, May 2025*

To achieve these objectives, our first action would be to create a Carbon Neutral Campus Steering Committee responsible for making initial decisions about the charrette.

The members of the Steering Committee will work with Summers & Mondragon, an urban planning firm with experience with issues of renewable energy and facilitating charrettes. Summers & Mondragon will also help us meet LEED standards (Leadership in Energy and Environmental Design) developed by the U.S. Green Building Council (http://www.usgbc.org).

The purpose of the Steering Committee would be to lay a solid foundation for the charrette. The members of the Steering Committee would include a range of people, including executive-level university administrators, faculty and staff, student leaders, and local citizens. We envision a planning committee of about twelve people meeting weekly for two months until the charrette is planned and scheduled.

Once the charrette is scheduled, the Steering Committee will write a report to the Tempest Foundation and the university president that describes its actions and decisions. The Steering Committee would invite feedback from the Tempest Foundation on this report.

Step Two: *Create a Charrette Library, Website, and Social Media Presence July 2025*

The Steering Committee will work with Summers & Mondragon to assemble information about the campus and identify options for renewable energy, conservation, and public transportation.

Working with librarians from Durango University, Summers & Mondragon will create a physical and virtual library that brings together any information that might be useful during the charrette. That way, participants in the charrette would already have the necessary information. The library would include documents, books, information from online sources, archival materials, and any other data participants might need. Durango University's librarians, led by Gina Sanders, would set aside a separate room for these materials and organize them into an accessible, cross-referenced, and electronically searchable system.

With the charrette library in place, the urban planning firm will then work with Durango University's communication officers to develop digital spaces that offer information and updates on the Carbon Neutral Campus Project, as well as provide opportunities for public participation. Documents from the Carbon Neu Campus library would be made available to the public through this digital platform and social media. Upda about the Carbon Neutral Campus Project will be accessible through a link on Durango University's homepage.

When this step is concluded, we will have developed a library of materials that can be accessed locally and through the Internet. The Carbon Neutral Campus digital space will also serve as a forum for public comme and an information clearinghouse.

Step Three: Host the Charette at the Student Union, September 2025
In Fall 2025, we will host a weekend charrette that brings together stakeholders interested in participating i the planning process. We expect about 100 people to attend, so we will reserve the Student Union's Chandl Ballroom and breakout rooms for the weekend. We will also order lunches and refreshments, so participant the charrette can stay focused on the planning process.

At the Friday evening kickoff meeting, we will introduce the facilitators from Summers & Mondragon. The facilitators will explain the Carbon Neutral Campus Project, discuss how the charrette will work, and make Carbon Neutral Campus library available to participants. We will then divide into design teams of six participants, allowing team members time to introduce themselves to each other.

On Saturday, each design team will develop its own plan for converting the campus to renewable energy sources. Experts from Summers & Mondragon and the Environmental Engineering Department will work v each team to explain technological abilities and limitations. They will also answer any questions that might arise in the design teams. The experts, however, will only serve as resources for the teams. Our aim is to maximize the creativity of the design teams by offering guidance without limiting their ability to be innova

Each team will have two hours on Sunday afternoon to finish its plan and create a slide deck. Then, each te will present its plan to the assembly. Members of the audience will be allowed to ask questions, identify ea plan's strengths, and probe any weaknesses. The Sunday meeting will be recorded, and the facilitators from Summers & Mondragon will take close notes on the proceedings. At the end of this meeting, all the plans w be submitted by the design teams,

Our expectation is that the design teams will develop plans that incorporate a variety of renewable energy sources, such as wind, solar, and geothermal. We also expect them to offer ideas for conserving energy and improving public transportation, such as introducing carbon-neutral buses (Figure 5). The design teams wil asked to develop plans that have both long-range and short-term features: (1) a long-range plan that elimina or offsets all greenhouse gases produced on campus and (2) a short-term plan that allows us to make immediate changes that will help us conserve energy and reduce our emissions of greenhouse gases.

Ultimately, the charrette will aim to draw on the collective creativity of the participants. Charrettes used for urban planning have been shown to bring out more creativity and knowledge than would be gathered by an urban planning firm alone. Moreover, charrettes like this one bring more stakeholders into the planning process, encouraging increased buy-in and less resistance to change. The community participates in the planning process, fostering a sense of identity and cooperation.

When this step is completed, we will put copies of the design and a video recording of the Sunday meeting online. We will also write a progress report for the Tempest Foundation that highlights the weekend's events and post the report on social media.

Renewable Energy Laboratory (NREL)

Step Four: Presentation of the Strategic Plan, October 2025

Using the plans from the charrette, Summers & Mondragon will develop a comprehensive Carbon Neutral Campus Strategic Plan for converting the campus to renewable energy. They will also identify any limitations that might keep us from achieving the goals discussed in the charrette.

At a Saturday meeting one month after the charrette, Summers & Mondragon will present the draft of their Carbon Neutral Campus Strategic Plan to the charrette participants. They will explain their version of the plan and solicit feedback from the audience. These proceedings will be recorded and posted online.

Our goal for this meeting will be to reach a consensus among stakeholders. If the charrette process is successful, the participants in the charrette will rally around the plan because they helped create it. When this meeting is over, we will put a copy of the design and a recording of the meeting on the Carbon Neutral Campus digital space. We will write another progress report to the Tempest Foundation that illustrates and discusses the plan developed by Summers & Mondragon.

Step Five: Finalizing the Plan, December 2025

Using the comments from the meeting, Summers & Mondragon will develop a final version of the Carbon Neutral Campus Strategic Plan. The full version will be sent to President Wilson within two months. Then, the final plan will be submitted by President Wilson to the university's Board of Regents for consideration.

The Carbon Neutral Campus Strategic Plan will provide a blueprint for converting the campus to renewable energy sources while minimizing the campus's emissions of greenhouse gases and other pollutants. Upon approval by the Board of Regents, the Carbon Neutral Campus Strategic Plan will guide all future decisions about building and renovating the campus. All campus budgeting, construction, and renovation decisions will be required to satisfy the Carbon Neutral Campus Strategic Plan guidelines.

We will present the Carbon Neutral Campus Strategic Plan to the Tempest Foundation at its January 2026 meeting. At that point, we can answer any questions about our implementation plans.

Dissemination

One of our goals is to establish a path that other universities can follow. For this reason, we will disseminate our plan through various venues, including the Carbon Neutral Campus digital space, the university website, national academic conferences, social media, and a variety of scholarly and popular publications. We will make the Carbon Neutral Campus Strategic Plan available to anyone who requests it. That way, other universities can use it to help themselves host their own charrettes and develop their own strategic plans. Meanwhile, at conferences, our administrators and faculty will present the results of the charrette. These conference presentations will lead to publications in academic journals and popular magazines.

The Tempest Foundation will be prominently mentioned on our website and in any printed materials related to this project. At conferences and in articles, the Tempest Foundation will be warmly thanked for supporting this project.

Assessment

To assess the program, we will retain two outside evaluators who are urban planning and renewable energy experts. We will submit their credentials for consideration and approval by the Tempest Foundation. Funds from the grant will be used to pay their expenses and an honorarium of one thousand dollars each.

The evaluators will observe all aspects of the Carbon Neutral Campus Project and have full access to any participants, meetings, and materials. When the project is completed, the evaluators will write a report to the Tempest Foundation that discusses their impressions and their appraisal of our efforts.

Qualifications of the Carbon Neutral Campus Team

Durango University is an ideal setting to undertake this revolutionary transformation to renewable energy. Located in southwestern Colorado, our campus has access to alternative renewable energy forms, including solar, wind, and geothermal energy. We also have a forward-thinking administration committed to converting the campus to renewable energy and achieving carbon neutrality. We want to be a positive role model for other universities in the United States.

Biographies of Key Personnel

The project leaders for the Carbon Neutral Campus Project will be Professor George Tillman and Vice President Anne Hinton. Key support will be provided by Diane Smith, a partner at Summers & Mondragon, and Gina Sanders, a Durango University research librarian.

- **Professor George Tillman, PhD,** is the John Connell Chair of Environmental Engineering. He has worked on geothermal energy issues for twenty-two years. He has authored or co-authored fifty-four articles on renewable energy and has worked with several small towns in Colorado to incorporate renewable energy into their energy grids. In 2013, Dr. Tillman was awarded the Environmental Engineer of the Year Award by the Colorado Environmental Protection Agency. He has been a principal investigator on $6.2M of grants in renewable energy research.

- **Anne Hinton, PhD,** is the vice president for Physical Facilities at Durango University. Before taking this position, she was dean of the College of Management. She is a specialist in international management and entrepreneurship. Her most recent book, *Managing Complex Change: A Guide to Entrepreneurship*, was published in 2015. She is a leading expert in fostering creativity in diverse teams of people.

- **Diane Smith, MUP,** is a partner with Summers & Mondragon, an urban planning firm located in Santa Fe, New Mexico. Smith has facilitated twenty-three urban planning charrettes and has worked with numerous cities on urban renewal, including Albuquerque, Chicago, Toledo, Santa Fe, and Santa Barbara. Summers & Mondragon has been a leading firm in using "New Urbanism" to reconceptualize commerce and transportation and make cities more pedestrian-friendly and less congested.

- **Gina Sanders, MLS,** is a senior research librarian at the Laura Vasquez Library on the Durango University campus. She has worked with many departments and local groups to gather information on technology issues. Her specialty is assembling information into electronically accessible formats. Her work with the Durango University Digital Archives won a Top Innovator Award from the National Librarians Council.

Curriculum vitae for the project leaders are included with this grant application.

Other faculty and staff members will also be assigned to this project. Durango University is the home of one of the leading environmental engineering departments in the nation. Our five faculty members and their graduate students will serve as resources for the community and the charrette. Meanwhile, staff members will be involved in developing the website for the charrette and assembling the library of documents and materials.

Background on Durango University

Another strength is Durango University itself. Since its founding in 1912, the university has had a long history of environmental leadership in the region. It was one of the first universities in the Southwest to develop an environmental engineering program. The program began in our Mining Engineering Department in 1958, drawing leaders in environmental engineering to its faculty. In 1980, the department was renamed Environmental Engineering to reflect environmental conservation as its primary mission.

Past Grants at Durango University

The university has received various grants from government sources and foundations. Recently, the National Science Foundation awarded the Environmental Engineering Department a $2.2 million grant to research opportunities for using geothermal energy in remote mountain communities. A $3.3 million Department of Energy grant was awarded to the Electrical Engineering Department for research into solar energy. The U.S. Department of Education also awarded Durango University a $780,000 grant to develop an educational program on the climate crisis that could be used in high schools throughout the United States. We have received funding from private foundations for our efforts to address world poverty in mountainous areas worldwide. The Geneva Foundation provided a $930,000 grant to one of our research teams to help improve access to electricity for people in the Himalayas.

Durango University has a track record of success in environmental issues. To learn more about the university and its successful grant-funded projects, please visit http://www.DurangoU.edu and http://www.DurangoU.edu/research.

Conclusion: The Benefits of the Carbon Neutral Campus Project

Let us conclude with a discussion of the costs and benefits of the Carbon Neutral Campus Project. We request $54,530 from the Tempest Foundation to help us develop the Carbon Neutral Campus Strategic Plan. With this strategic plan in place, we can start taking positive steps toward strengthening our community and addressing the causes of the climate crisis on our campus. Our itemized budget is enclosed, and we believe that cost-sharing with the Tempest Foundation is an integral part of our contribution toward the project.

The benefits of the Tempest Foundation's support for the Carbon Neutral Campus Project will be well worth the investment:

- The Carbon Neutral Campus Project will allow us to begin converting our campus to renewable energy sources from the grassroots up. The charrette will draw from the community's knowledge, wisdom, desires, and experiences, while encouraging people to participate in the transformation process.

- The Carbon Neutral Campus Project will generate and cultivate an ongoing dialogue about energy conservation and sustainable lifestyles, building unity and cooperation on our campus.

- The Carbon Neutral Campus Project will provide a prototype that other universities can follow to convert their campuses to renewable energy and a carbon- neutral status,

- The Carbon Neutral Campus library will be accessible worldwide through the Durango University website, offering a comprehensive resource on issues related to the climate crisis.

- The Carbon Neutral Campus Project will be supported by our Environmental Engineering program, which has a track record of success in energy-related projects.

The most significant benefit of the Tempest Foundation's investment will be the development of a comprehensive strategic plan that will guide the conversion of our campus to a carbon-neutral status. We can then centralize the Carbon Neutral Campus Project in our campus planning and take bold steps toward doing our part to solve the climate crisis.

We believe the Carbon Neutral Campus Project at Durango University will provide a way forward into the future—a future that is sustainable and environmentally sound. When the Carbon Neutral Campus Project is completed, Durango University will have demonstrated that energy independence, carbon neutrality, and sustainability are possible and highly advantageous for university campuses. Today, our society no longer has the luxury to wait for others to take the lead on the climate crisis. At Durango University, we believe we can take steps right now that will help all of us to prevent worst-case scenarios for the climate crisis.

Thank you for your time and consideration. We look forward to hearing from you about this request for a grant. If you have any questions, comments, or suggestions for improvement, please contact George Tillman, professor of Environmental Engineering, at (970) 555-0124 or Anne Hinton, vice president for Physical Facilities, at (970) 555-1823. We appreciate your willingness to consider our request.

References

Bailey, J., & Grossman, D. (2023). Getting transportation right: Ranking the states in light of new federal funding. Retrieved from https://www.nrdc.org/sites/default/files/2023-11/state-transportation-ranking-federal-funding-report.pdf

Intergovernmental Panel on Climate Change (IPCC). (2023). *AR6 synthesis report: Climate change 2023*. Retrieved from https://www.ipcc.ch/report/ar6/syr/

Philips, G. (2018). *A history of Durango University: From mountain college to world-class university*. Durango, CO: Durango University Press.

U.S. Green Building Council. (n.d.). *Home page*. Retrieved from http://www.usgbc.org.

APPENDIX B: FUSIONFACTORY PROPOSAL

The FusionFactory

**Proposal to
The Riverview Heights Chamber of Commerce**

Blue Lion Ventures

Appendix B: FusionFactory Proposal

Executive Summary

Riverview Heights, Illinois, holds immense potential as a sleeping economic giant. Despite its advantageous location and size, Riverview Heights lags in high-tech investment and talent attraction, with its downtown and the Ennova Innovation District notably underdeveloped. This proposal by Blue Lion Ventures introduces "The FusionFactory," a downtown coworking space designed to invigorate Riverview Heights's technology business sector and catalyze economic growth.

The FusionFactory aims to be a state-of-the-art coworking space, providing a nexus for technology startups and an innovation hub in Riverview Heights. Riverview Heights currently lacks specialized coworking facilities, and this initiative is poised to fill a significant gap. The space will offer resources such as networking, mentorship, and investor access, fostering small business growth and drawing attention to Riverview Heights as a technology hub.

Riverview Heights's small business development efforts have not taken off, particularly in the technology and entrepreneurial sectors, due to economic barriers and scarce office space. A lack of anchoring hubs in Riverview Heights's entrepreneurial ecosystem and outdated economic and technological infrastructure exacerbate these challenges. Additionally, Riverview Heights must bolster cultural, social, and material factors to attract the modern tech workforce.

Blue Lion Ventures proposes a three-phased approach:

Phase I: Re-convene the Ennova Innovation District Coalition, redefining goals and developing a strategic plan focused on creating a network of entrepreneurial hubs.

Phase II: Renovate a downtown warehouse into The FusionFactory, emphasizing sustainable and modern design with high-tech infrastructure.

Phase III: Build an integrated entrepreneurial ecosystem, leveraging resources from local hubs like universities, business incubators, and the Chamber of Commerce.

With over 25 years of experience in innovative solutions, Blue Lion Ventures boasts a strong team led by CEO Jacob Johannsen and COO Stephanie Anderson, alongside a highly skilled workforce. Their history of successful coworking space designs and community-oriented approaches makes them uniquely suited for this project.

This proposal outlines substantial benefits for Riverview Heights, including reduced startup costs, enhanced small business infrastructure, and preservation of Riverview Heights's charm. With a projected cost of $250,000, the investment promises high returns in terms of economic growth and small business development. The plan also aims to retain local university talent and enhance the city's appeal to entrepreneurs.

Blue Lion Ventures presents a comprehensive, multi-phased strategy to transform Riverview Heights into a thriving hub for small businesses and innovation, inviting further collaboration and input from local stakeholders.

The FusionFactory Co-Working Space

Riverview Heights, Illinois, is a sleeping economic giant in the greater Chicago area. Located in the beautiful Fox River valley, Riverview Heights is the fourth-largest city in Illinois. It's also one of the charter members of the Illinois Technology and Research Corridor, which includes affluent communities like Naperville, Downers Grove, Oak Brook, and Hinsdale. Riverview Heights's immediate neighbors are highly sought-after smaller towns like St. Charles, Geneva, and Batavia.

And yet, Riverview Heights has been behind the curve in attracting the high-tech investment and talent that is stimulating growth in neighboring communities. Even after years of development efforts, many of Riverview Heights's downtown buildings are vacant, and the once-promising Ennova Innovation District seems to be stalled due to the COVID-19 pandemic. In other words, Riverview Heights has been running in place, despite the growing economies around it.

Here at Blue Lion Ventures, a Chicago-based investment firm, we strongly believe that Riverview Heights has the potential to be a technology center with a bright future.

This proposal describes a plan to launch The FusionFactory, a state-of-the-art downtown coworking space tailored to the needs of small technology businesses. The FusionFactory will serve as an innovation hub that can anchor an entrepreneurial ecosystem, which attracts and nurtures tech startup businesses in Riverview Heights. It will also serve as a conduit for bringing investments and economic growth into Riverview Heights. Coworking spaces have emerged as pivotal platforms for small businesses to collaborate and grow, especially in the technology sector. Surprisingly, despite Riverview Heights's size and optimal location, it does not have any coworking spaces, no less one that serves the unique needs of high-tech startups. The FusionFactory, as a multimodal coworking space, will not only provide a physical location for work but will also offer vital resources such as networking opportunities, mentorship programs, and access to investors.

To attract and nurture small technology businesses, we would like to partner with Riverview Heights's leaders to grow the local economy, create jobs, and attract investment. This initiative will also draw attention to Riverview Heights as an up-and-coming tech hub, inviting further investment and development in the region. The coworking space will feature state-of-the-art facilities: high-speed internet, ergonomic workstations, private meeting rooms, and collaborative open areas. Additionally, we plan to host regular events, workshops, and seminars, led by industry experts, to nurture the growth and development of our members.

To realize this vision, we seek the collaboration and support of the mayor's office, the Riverview Heights Chamber of Commerce, and investors within the Greater Chicagoland area. Your expertise, resources, and networks are invaluable in making this project a success. Together, we can create a landmark initiative that will not only benefit small businesses but also help Riverview Heights achieve its economic potential.

The Opportunity

Before describing our plan, let us first highlight the opportunity for a co-working space by analyzing the reasons why small businesses struggle to succeed in Riverview Heights. The city has been able to attract several relocated businesses, including a car parts manufacturer, two meat processing plants, and a custom plastics plant. Unfortunately, the small business sector has not kept pace with the manufacturing sector. Currently, Riverview Heights ranks in the 50th percentile for small business development rates in the state.[1] The lack of small businesses and the shortage of entrepreneurial

---

[1] Illinois Economic Growth Office. (2024). *Illinois Data Bank.* Springfield, IL: State of Illinois.

ventures has limited Riverview Heights's ability to attract new residents and retain recent graduates from nearby colleges and universities. Broadly, the stagnation of small business development in the city is due to two related causes: (1) a lack of office and light manufacturing spaces and (2) the economic barriers to starting a business in town.

In this analysis, we discuss why Riverview Heights, despite its many advantages, isn't already an economic and innovation powerhouse. Riverview Heights has many advantages, but a few important factors have been holding the city back from becoming a destination for entrepreneurs, start-ups, and tech companies that want to relocate. So, before offering our plan, we want to explain why Riverview Heights may be struggling to build a strong entrepreneurial ecosystem.

Hubs as Anchors within an Entrepreneurial Ecosystem

One factor is the absence of facilities that anchor Riverview Heights's entrepreneurial ecosystem. A successful technology ecosystem needs to be built from the ground up, rather than imposed from the top down. Successful entrepreneurial ecosystems typically include a network of coworking spaces, universities, small-business incubators, business accelerators, investment firms, and government leadership. These hubs provide a supportive network for emerging companies. In the absence of this critical mass, small businesses will struggle to gain traction because of the high costs they bear for facilities, materials, services, transportation, and labor, especially when competing with other more established business ecosystems.

Developing a critical mass of hubs and small businesses is vital. A case in point is the Riverview Heights's Ennova Innovation District, which looked like a promising step toward building such an entrepreneurial ecosystem. The Ennova Innovation District may have been the right idea at the wrong time because its launch in January 2020 coincided with the onset of the COVID-19 pandemic. As the pandemic swept through the area, the cultural, social, and material attributes needed to build an entrepreneurial ecosystem were put on hold or never developed. By the time the pandemic was over, the momentum behind the Ennova Innovation District had all but disappeared. Moreover, the district was rather top down in structure. The sponsors were primarily big corporations rather than the hubs and local businesses that would ultimately form a sustainable entrepreneurial ecosystem. When those sponsors had to shift their attention to navigate the pandemic, they quickly forgot about the Ennova Innovation District

The drive to establish the Ennova Innovation District was correct, and the support from major corporations and politicians provided a significant amount of momentum. However, building a sustainable entrepreneurial ecosystem requires a bottom-up approach, anchored by interconnected hubs that can nurture growth over time and are resilient in the face of unexpected exigencies.

Economic and Technological Infrastructure

Like many cities built for heavy manufacturing in the early 20th century, Riverview Heights is saddled with economic and technological infrastructure developed for a bygone era. It's not enough to layer a digital infrastructure over an existing heavy manufacturing infrastructure. Instead, a digital-forward strategic plan is needed that sets milestones for sustainability, transportation, and digital innovation. Right now, the Ennova Innovation District does not have a solid governance structure that can coordinate and facilitate the execution and oversight of "smart city" projects that align with the strategic plan.

Essentially, Riverview Heights needs to recognize that it is competing with more established technology ecosystems in neighboring communities such as Naperville, Downers Grove, and Oak Brook that already have these economic and technological infrastructures in place. To compete, Riverview Heights needs to lean into core strengths that are unique to Riverview Heights, such as lower labor costs, inexpensive corporate real estate, attractive tax rates, as well as organizations like Invest Riverview Heights and other public-private partnerships. By combining these elements, Riverview Heights can transition from its manufacturing roots to a modern, efficient, and citizen-focused smart city. The fact that a significant building like the former Robert Morris University campus was vacant for nearly three years before being repurposed for a tech company indicates potential gaps in available infrastructure suited to the needs of tech-oriented businesses.

A successful entrepreneurial ecosystem in Riverview Heights calls for a critical addition to its infrastructure—a dedicated space for creativity, collaboration, and business growth.

Cultural, Social, and Material Factors

As Professor Lora Gellar, an expert in entrepreneurial ecosystems, suggests, the success of a region's entrepreneurial ecosystem is deeply intertwined with the interplay of cultural, social, and material attributes.[2] Gellar points out that if these cultural, social, and material attributes are lacking or their connections are weak, the region may face difficulties in attracting startups and fostering innovation.

Today, tech workers can live and work from just about anywhere. Miami, Libson, Dublin, Wrigleyville. So, what might attract them to a place like Riverview Heights? This city has many advantages, including inexpensive housing, a diverse population, low-cost labor, regional universities, quality healthcare, and Chicago nearby (but not too near). In other words, Riverview Heights has all the material attributes that tend to attract entrepreneurs and cutting-edge businesses. However, it could have more of the cultural and social amenities that young and creative professionals look for, such as high-end restaurants, green spaces, a vibrant arts and music scene, recreational facilities, and a tech-friendly infrastructure.

Quality of life is a major factor in attracting creative people and sustaining entrepreneurial ecosystems. Riverview Heights has many of the attributes of these ecosystems, but more effort could be put into improving this aspect of the community.

Effects of Inaction

If left unattended, the current issues hindering small business growth in Riverview Heights will not go away. The city will continue to experience low small business growth rates and continue to be over-reliant on the manufacturing sector. Moreover, the city is in direct competition with other cities in the region for these businesses and educated workers. Without taking steps to resolve these issues, Riverview Heights will not be able to compete for these businesses and workers. Likewise, if no action is taken then the downtown area will continue to decay, resulting in greater expenses in the future. Ultimately, we could see the rate of Riverview Heights University students that stay in the region fall below the already low rate of 5%. These demographic trends could further hamper the small business growth in the city as it would force the small businesses into increased competition for employees, driving up initial development costs.

Our Plan: The FusionFactory, a Multimodal Coworking Space

Helping Riverview Heights increase its number of small businesses and attract innovators to the region requires a plan that allows the town to capitalize on its current strengths and resources without

---

[2] L. Gellar. (2024). How to build a better entrepreneurial ecosystem. *Journal of Entrepreneurial Growth*, 3(1), 231–245.

major expenditures or disruptions to the local manufacturing concerns. We believe that the plan must meet the following objectives:

- Lower startup costs and overhead costs for new small businesses in town
- Make use of the current infrastructure in town
- Offer a solution that will appeal to current small business owners, aspiring small business owners, recent university graduates, and out-of-town investors and entrepreneurs

To meet these objectives, Blue Lion Ventures has developed a plan that will help Riverview Heights meet its development goals. The plan entails converting older warehouse space in Riverview Heights's downtown district into coworking spaces. Coworking spaces will lower startup costs and minimize overhead expenditures for new or aspiring small business owners—especially in the technology and professional services sectors. Moreover, the number of coworking spaces can be increased to meet the demands of new businesses. Previous coworking spaces developed by Blue Lion Ventures have doubled and tripled in size to accommodate growing demand.

Our plan will be implemented in four major phases. First, we will assess the community's needs and work to ensure buy-in from current business owners. Second, we will build a coworking space in Riverview Heights by renovating a warehouse in the downtown district. Third, we will conduct a marketing campaign to build further interest in the space and recruit tenants. Fourth, we will evaluate interest and engagement with the coworking space one year after it opens.

Phase I: Re-Convene the Ennova Innovation District Coalition of Partners

Our experience has shown that the success of any community endeavor depends on overall community investment in the project. To develop community engagement with the coworking space, we will begin by reconvening the partners who launched the Ennova Innovation District, including local government officials, business leaders, educators, and community representatives. At a meeting, we would stress the importance of building the entrepreneurial ecosystem from the ground up. To revitalize partnerships and adapt to the evolving landscape, we propose the following strategic steps:

- **Re-establishing goals and objectives, clearly defining the goals and objectives for reconvening the partnership**. This step will consider the changing post-COVID pandemic needs of Riverview Heights, emerging technological trends, and the lessons learned from past experiences.

- **Developing a strategic plan that stresses the importance of building a network of hubs including the FusionFactory.** The strategic plan would also bring in and formalize relationships among network hubs, such as local universities, the Riverview Heights Chamber of Commerce, investment firms, and the City of Riverview Heights's city services. We would lay the groundwork for launching small-business incubators and business accelerators.

- **Hosting a kick-off event at which the Ennova Innovation District will be relaunched with the aim of building excitement, encouraging collaboration, and promoting partnership.** We would stress a bottom-up development approach, emphasizing launching a coworking space, strengthening the relationships among hubs, and supporting small-business development in downtown Riverview Heights.

At the end of Phase I, we will write a progress report addressed to the City of Riverview Heights and the Riverview Heights Chamber of Commerce that assesses our progress and updates our objectives and strategic plan.

Phase II: Renovate a Downtown Warehouse into the FusionFactory Co-Working Space

Establishing a coworking studio in a renovated warehouse in downtown Riverview Heights, Illinois involves five key steps:

- **Conduct a thorough assessment of the selected warehouse, including its structural integrity, existing utilities, and space potential.** Then, we will develop a detailed business plan that outlines the vision, target market, financial projections, and unique selling points of the coworking studio. This plan will emphasize how the studio will contribute to Riverview Heights's economic growth and community development.

- **Collaborate with architects and interior designers to create a layout that maximizes the space, ensuring areas for individual workstations, meeting rooms, communal areas, and amenities.** We seek to renovate the warehouse with a focus on sustainability and modern design, preserving historical elements where possible to maintain the character of downtown Riverview Heights.

- **Install high-speed internet and modern information technology infrastructure to support the diverse needs of future tenants, including freelancers, startups, and remote workers.** This will ensure that the space is equipped with state-of-the-art office equipment and facilities, like video conferencing rooms, ergonomic workstations, and smart lighting systems.

- **Develop a marketing strategy to attract a diverse range of tenants, leveraging local networks, social media, and partnerships with local businesses and educational institutions.** We will plan events and workshops for community engagement, positioning the coworking studio as a hub for innovation and collaboration in Riverview Heights.

- **Launch a grand opening event to generate buzz and formally introduce the coworking space to the Riverview Heights community.** We will establish an operational plan for managing the space, including staffing, membership structures, and ongoing maintenance, ensuring a vibrant, productive, and collaborative environment for all users.

Phase II of our plan will establish the FusionFactory, which will foster a dynamic community of professionals, contributing significantly to the economic and cultural fabric of Riverview Heights, Illinois.

Phase III: Build an Entrepreneurial Ecosystem

The third phase entails building an entrepreneurial ecosystem in Riverview Heights, Illinois, with key hubs like a coworking space, local universities, small business incubators, the Chamber of Commerce, and city government. This phase will include four steps:

- **Establish a coalition among the major hubs that will forge a shared vision and objectives for the entrepreneurial ecosystem, ensuring alignment of goals and resources.** Toward this end, the coalition would create a formal agreement or memorandum of understanding that outlines the roles and responsibilities of each entity in supporting and nurturing the ecosystem.

- **Integrate resources and build a knowledge base that leverages the unique strengths of each hub.** For example, the university can provide research and development support, the small business incubator can offer mentorship and business development programs, the coworking space can serve as a networking and collaborative platform, the Chamber of Commerce could offer leadership and mentorship, and the city government can facilitate regulatory support and access to funding. As a member of the coalition, the FusionFactory can form integrated programs and initiatives that allow for resource sharing, such as joint workshops, speaker series, and mentorship programs, to foster a culture of innovation and collaboration.
- **Develop networks by engaging with the broader Riverview Heights community, including local businesses, investors, and entrepreneurs, to build a supportive network.** This can involve organizing networking events, community forums, and innovation challenges. We will develop communication channels, like a dedicated website or social media platforms, to keep the community informed and engaged with the ecosystem's activities and opportunities.
- **Create a plan for sustainability and growth that sets metrics to evaluate the impact and success of the ecosystem.** The plan will include metrics such as the number of startups incubated, jobs created, or investments attracted. The coalition can develop for long-term sustainability by securing diverse funding sources, including grants, sponsorships, and public-private partnerships.

In Phase III, we will help Riverview Heights and the broader coalition of entrepreneurial hubs establish a robust entrepreneurial ecosystem that fosters innovation, supports startups, and drives economic growth. This ecosystem will leverage the combined strengths and resources of its key hubs.

Qualifications

At Blue Lion Ventures, we know this is a pivotal time for Riverview Heights. To foster small business growth and attract entrepreneurs to the region, Riverview Heights needs a firm that has experience developing and managing innovative projects. Blue Lion Ventures is uniquely qualified to help Riverview Heights because we have three decades of experience providing clients with innovative solutions that are tailored to their needs.

Management and Labor

With more than fifty combined years in business consulting, our management team offers the insight and responsiveness you require to handle Riverview Heights's complex needs.

Jacob Johannsen, our CEO, founded Blue Lion Ventures in 1995. He completed his Ph.D. in organizational behavior at the Harvard Business School and spent a decade working for IBM as a systems analyst. Under his leadership, Blue Lion Ventures has provided innovative strategies for dozens of local governments, nonprofit organizations, and businesses, including several Fortune 500 companies. Since its founding, Blue Lion Ventures has grown from a six-person operation into a nationally recognized firm with more than 50 specialized employees.

Stephanie Anderson, our COO, was one of the original employees of Blue Lion Ventures and has been with the company for over 25 years. She began her career with Madison & Company, serving as a project manager for dozens of national and international endeavors. Her expertise in organizational development and her commitment to fostering a culture of innovation and excellence have been instrumental in shaping Blue Lion Venture's growth trajectory and building its reputation as an industry leader.

Mohammed Fahimi, our lead designer for this project, earned his Master of Architecture from Cornell University in 2011. Mohammed has designed over a dozen coworking spaces in the past and specializes in designing spaces that are both functional and aesthetically pleasing, with a focus on fostering collaboration and creativity. His expertise in utilizing natural light and sustainable materials has been instrumental in creating eco-friendly and inviting work environments.

The resumes of our management team are included in Appendix B.

Our management team is supported by a forward-thinking corps of high-technology and business-savvy employees. Blue Lion Ventures employs over 25 specialized employees, whose expertise includes business management, interior design, and systems analysis. Blue Lion Ventures recruits its employees from the most advanced universities in the United States, including Stanford, MIT, Illinois, Iowa State, Purdue, New Mexico, and Carnegie Mellon.

Corporate History

Since 1995, Blue Lion Ventures has provided cutting-edge solutions for small businesses, large corporations, local governments, and nonprofit organizations. Blue Lion Ventures has been called an "industry leader" by *Business Outlook*, a "perfect mixture of adaptiveness and experience" by *Consultant Quarterly*, and "the specialists you've been waiting for" by *Small Business Strategies*. Blue Lion Ventures's award-winning office space is modeled on coworking spaces. We know the advantages of cutting-edge coworking spaces designed to foster creativity and collaboration because we work in one every day.

Experience You Can Trust

As team members of a dynamic growing business that is well acquainted with the issues faced by small businesses and entrepreneurs, we at Blue Lion Ventures understand the unique challenges that face Riverview Heights. The keys to success are innovation, flexibility, and experience. Blue Lion Ventures will help Riverview Heights transform into a thriving small business and technology hub.

Plan Benefits and Project Costs

Let us summarize the advantages of our plan and discuss the costs. Our preliminary research shows that Riverview Heights will continue to be a regional manufacturing hub in the coming decades. Coupling Riverview Heights's success in manufacturing with a robust small business community offers a way for Riverview Heights to become a major economic hub in the region. At Blue Lion Ventures, we believe that a coworking space will help Riverview Heights foster small business growth, attract technological and entrepreneurial businesses, and retain its small-city charm.

Cost is the most significant advantage of our plan. As shown in Appendix A, implementing our plan would cost approximately $550,000, of which we would raise $300,000 from investors. We believe a

$250,000 investment by Riverview Heights in its small business infrastructure will pay for itself within 5 years. Our plan has additional benefits that go beyond simple costs:

- *A coworking space will significantly decrease the barriers to small business development in the city, especially for entrepreneurial and technological ventures.* When a new small business in Riverview Heights reaches the point of requiring office space, the owner can simply register with the FusionFactory. For a small monthly fee (approximately 10% of the cost of renting an office suite), the business owner will have access to the latest advancements in office space design. The business owner can focus on their business rather than on finding a space in which to work.
- *A coworking space will help Riverview Heights retain Riverview Heights University Students following graduation.* New business and entrepreneurial infrastructure will help Riverview Heights retain a greater share of Riverview Heights University students following graduation because the city will have jobs and opportunities to offer recent graduates. Moreover, by increasing the number of university graduates who remain in the city, Riverview Heights will increase the overall education levels of its workers and generate new markets for future small businesses.
- *A coworking space will help preserve Riverview Heights's small-city charm.* By renovating a warehouse rather than building a new structure, we will ensure that the coworking space will blend seamlessly in with Riverview Heights's small-city charm. The space will speak to Riverview Heights's history as an agricultural hub, its present as a manufacturing center, and to its future as the home of many small businesses and entrepreneurial endeavors. This project may receive additional support from community members who have a personal stake in revitalizing the downtown district.

With a coworking space, Riverview Heights will have a small business infrastructure that can compete with that of major cities while also offering lower costs of living and lower costs of market entry for entrepreneurs.

Thank you for giving Blue Lion Ventures the opportunity to work with you on this project. We look forward to submitting a full proposal that describes our plan in greater detail. Our COO, Stephanie Anderson, will contact you on December 15 to discuss this pre-proposal with you.

If you have any suggestions for improving our plan or you would like further information about our services, please call Mohammed Fahimi, our lead designer for this project, at (789) 555-0164. Or, you can contact him at mfahimi@bluelion_consulting.com.

ABOUT THE AUTHORS

Richard Johnson-Sheehan is a Professor of Rhetoric and Professional Writing at Purdue University. He researches and publishes on communications in science, technology, entrepreneurship, and healthcare. He is also the President of Phronesis, LLC, a communications and consulting company that produces and edits books, articles, and manuscripts in fields related to science, technology, entrepreneurship, and medicine. He specializes in technical proposals, grants, and business models for start-ups. Johnson-Sheehan has collaborated with and consulted for a broad range of enterprise and small businesses and organizations, including Intel, the Indiana Department of Transportation, Sumitomo, the United Way, Sandia National Labs, Lockheed, University of New Mexico Hospital, Pearson Educational, Integrated Robotics, among others. Johnson-Sheehan is the author of several popular college textbooks, including *Technical Communication Today* (7e), *Writing Today* (5e), *Argument Today* (2e), and *Technical Communication Strategies for Today* (3e), as well as this book, *Writing Proposals and Grants* (3e). He earned his PhD in Rhetoric and Professional Communication from Iowa State University. He lives with his wife, Tracey, and his children in West Lafayette, Indiana.

Paul Thompson Hunter is a PhD Candidate in Rhetoric and Composition at Purdue University, where he received the Janice Lauer Dissertation Award and the Crews and Biggs Dissertation Fellowship. He received his MA in Technical and Professional Writing from the University of North Carolina at Charlotte. His work has appeared in *IEEE Transactions on Professional Communication*, the *Journal of Technical Writing and Communication*, and *Communication Design Quarterly*. He teaches technical writing, business writing, and scientific communication. His current research examines the intersections of entrepreneurship, artificial intelligence, and user experience design. Since the age of fourteen, Paul has worked as a file clerk, a digital archivist, a copywriter, an editor, and a communications consultant. He has experience in authoring, revising, and reviewing business proposals, research grants, and funding applications for small businesses, nonprofit organizations, and academic institutions.

INDEX

access points, 168, 172, 174, 181, 184
accountant: importance of, 37, 123, 125, 127, 130–133, 136, 137
alignment: in page design, 173; subject, 163–164; using lists, 173, 177
analogies: using, 161
appendices: use and placement, 42, 197
artificial intelligence (AI), xiii–xiv, 8, 10, 33, 35, 38–39, 43, 46, 64, 69, 79, 81–83, 88, 92, 101, 107, 111, 118–119, 122, 144–145, 163, 166, 184–185, 191–193, 200–202, 209; images, 192
assessment, 28, 74, 81, 84, 91, 100, 207; in project plan section, 84, 91
attitudes: of readers, xiii, 8, 39–40, 42, 45, 98, 114, 206

back matter, 197, 202–203
background information, 5, 14, 52, 54, 56, 61, 63, 111, 198, 208
Background section, 58, 60–63, 66, 68–75, 78–79, 83, 88–89, 94, 164–165, 187
balance: in expression, 4, 17; in page design, 144, 169, 171–172, 180
Benefits section, 60, 166; consequent, 113; direct, 113; hard, 113, 118, 120; soft, 113–114, 118, 121; value, 113, 115, 118, 120
bibliography: in proposals, 202, 204
biographies: of managers, 102–103, 107, 111, 203
body paragraphs, 70; described, 70
boilerplate, 107
borders: in page design, 170, 175, 179, 183
budget rationale, 137–139, 141, 202–203
budgets, 5, 117, 123, 126, 128–129, 134–135, 137–139, 141–143; fixed, 124–125; itemized, 56, 117, 123–124, 126, 130, 133, 139

character: use in argument, 98, 144, 200
charts: organizational, 103, 190
collaboration: on proposals, 96
concept mapping, xiii, 63–64, 108, 159,
conclusions, 56, 70, 112–120, 122, 137, 146, 150–151, 159, 198–199, 207
consistency: in page design, 169, 175, 178, 180, 208, 210
contexts: physical, 42, 178, 181
cost sharing, 136–137, 141
costs, 5, 11, 22, 24, 29, 40, 42, 56, 77, 100, 112, 115–120, 123–127, 130–133, 136, 137–141, 159, 187, 194, 196, 200–201; depreciation, 130; facilities and administrative (F&A), 136; fixed, 125; synopsis of travel, 131–132; variable, 125–126, 130, 139, 185, 187
costs and benefits, 5, 56, 112, 115–116, 119, 159, 200–201
cover pages, 181–182, 197, 200

deliverables, 23, 25, 87, 91, 113
describing, 103
design: of documents, Chapter 11 *passim*; 122, 168, 172, 179–180; principles of, 168
drawings, 24, 96, 178, 191, 193–194, 200

editing, xii, xiii, 8, 122, 144, 165, 179, 185, 193, 197, 205, 210
effects approach, 70–72
emotions, 158–159, 164; of readers, 158
equipment, 5, 66, 85, 99–100, 102, 104–105, 125–126, 130–131, 136, 138, 142
ethics, 40, 43, 162
ethos, 98, 108, 144
executive summaries, 200, 202
expression, 4

235

facilities, 5, 98–100, 102, 104–105, 107–108, 125, 130–131, 136–138, 157, 207; photographing, 183–184, 191, 193–194
facilities and administrative costs (F&A), 126, 136, 141, 143
fishbone diagram, 64, 66, 85–86; to develop project plan, 85; to find causes, 85
forecasting: using, 55, 58, 102, 208
formulas: using in proposals, 92, 117, 130, 202, 205
foundations funding, 29–30, 32, 47–50, 59–61, 94, 110, 120–122, 140–142, 181, 193, 210
front matter, 197

genre, 5–6, 10, 62
glossaries: in appendix, 203–204
goals, xii–xiii, 9–11, 14, 47, 72–73, 81, 83, 92, 94, 104–105, 153, 162
grabber, 53, 56, 60
Grants.gov, 13, 17, 27
graphics: bar charts, 186–187; Gantt charts, 92, 189; line graphs, 186; organizational charts, 103, 190; pie charts, 188; tables, 88, 96, 117, 137, 177, 179, 181, 183–184, 187–188, 193, 208
graphics, Chapter 12 *passim*; xiii, 8, 92, 122, 171–172, 175, 177, 179–185, 187, 189, 192–193, 196, 204, 208–210
grouping: in page design, 169, 175, 178, 180, 208

history: corporate or organizational, 62, 69, 100, 102–105, 108, 203

IEEE, 188, 204
in-kind contributions, 137
Internet: using, 5, 10, 33, 35, 39, 46, 59, 64, 68–69, 75, 78, 83, 88, 93, 101, 108, 120, 132, 139–140, 142, 164, 182, 191–194, 203, 209
interpretation, 4

introductions: to proposal, 45, 52, 54–56, 58–59, 61–62, 70, 75, 112, 159, 166, 197–198, 208
Ishikawa Fishbone Diagram, 64–66, 85, 86

labor, 65, 85, 99, 102–103, 124–127, 130, 139, 204; indirect, 130
lede, 53; *see also* grabber
letter of transmittal, 198
lists: nonsequential, 177, 179; sequential, 177
literature review, 63, 73
logic: examples, 152; reasoning, 152

main point: of proposal, 5, 59; of section, 70, 72
main points, 5, 52, 55–56, 58–60, 70, 72, 88, 101, 116, 119, 122, 138, 144, 172, 192, 198, 201–202, 208
management, 3, 7, 91–92, 99, 102–104, 108, 125, 127, 130, 139, 146, 150, 202, 204; budgeting for, 126–127; team, 102–103
matching funds, 136
materials, 24, 64–65, 68–69, 78–79, 82, 85, 92, 105, 107, 124–126, 131, 137–138, 140, 142, 197, 202, 204–205, 207; direct, 131; indirect, 131
memo of transmittal, 197–199, 209
metaphors: using, 161–162, 164, 166–167
methodology: in proposal, 24, 81, 92–93, 98
Methods section, Chapter 6 *passim*; 63, 72, 92, 93

National Endowment for the Humanities (NEH), 9
National Institutes of Health (NIH), 9, 36, 55, 124
National Science Foundation (NSF), 9, 15, 16, 23, 110, 124, 140, 141

Objectives: setting, 80–82, 86, 89; secondary, 80–83, 86–87, 89, 93
opening paragraphs, 70, 89, 105, 138, 166,

236

203, 208; appendix, 203; background, 73; project plan, 89, 91
opportunities: problems, 28; funding, 13, 18
organization of the proposal, 52, 174, 201

pace of writing, 164
paragraphs: plain, 151, 155–156
passive voice, 156–158
persona: creating, 105–108, 208
personnel: describing, 24, 64, 85, 98, 100, 102, 107, 202–203, 207
persuasion, 3–4, 10, 43, 158
philosophy: corporate, 102
photographs: using and placing, 183–184, 191, 193–194
Plan section, 80, 88–90, 93, 99
point of contact (POC), 25–26, 28–29, 31, 82
politics, 43
principal investigators, 126–127, 203
prior research, 63, 72–74
problems: defining, 20–25
process: design, 178, 180; writing, xiv, 8, 11, 12, 20, 36, 45, 93, 120, 197, 207–208
profits: in budgets, 39, 113, 127, 131, 133, 136
Project Plan section, Chapter 6 *passim*; 5, 60, 63, 72, 79–80, 82, 84, 86, 88–91, 93–99, 112–113, 118, 120, 163–164, 190
proposals: definition of, xii; external, 7, 125; formal, 118; grant, 3, 5, 9, 11, 15, 17, 23, 29, 32, 36, 39, 42, 46, 50, 59, 62–63, 67, 75, 87, 91, 96, 98, 109, 110, 122, 126, 136, 140, 161, 164, 181, 189, 193–194, 196, 198, 205, 209–210; implementation, 24–25, 113, 207; internal, 99; planning, 5, 24–25, 32, 82, 207; research, 23–24, 55, 72–74, 92–93, 207; sales, 24; unsolicited, 7
purpose, 5, 14, 33, 36, 45–47, 49, 51–52, 55–56, 58–59, 62, 70, 75, 82, 88–89, 91, 98, 102, 104, 122, 138, 151, 180, 190, 197–198, 201, 203, 206, 208–210; of proposal, 33; statement, 36, 55–56, 198, 206

Qualifications section, Chapter 7 *passim*, 60, 98–102, 105–113, 117–118, 120, 137, 190, 203
quality control, 24, 66, 85, 102

readers, types of: gatekeepers, 45; primary, 45; secondary, 45; tertiary, 37, 38, 50, 206
Request for Proposals (RFPs), 7, 13–18, 21, 25–30, 35, 46, 48, 60, 82, 93, 198
research proposals, 23–24, 72, 74, 92
resumes: placement of, 103, 203
reviewers, for grants, 27, 36, 39, 60–61, 74, 92–93, 122, 140, 181, 210
revision, 107, 147, 149, 154–155, 179, 206, 208–209
rhetoric, xii, 4, 10, 17, 20; definition of, xii; use of, 4, 10–11; visual, xiii, 68, 161–162, 166, 179, 184, 189, 191–192, 208
rhetorical situation, Chapter 3 *passim*; 17, 27, 33, 45–47, 51, 75, 178–181, 206, 209
rules, as design features: horizontal and vertical, 175

salaried, 126, 130
sentences, 33, 56, 58–59, 70, 105, 120, 144–145, 147–158, 162–167, 193, 198–199, 210; plain, 144, 158, 162; point, 153; support, 151–153; topic, 151–153, 165; transition, 151, 153, 165
similes: use of, 161
solutions, 77, 80, 83, 86, 96, 209
staff, 96, 103, 126–127, 130, 166, 181, 195; budgeting for, 117; description of, 103
stasis questions, xiv, 20, 25, 27–28, 31, 35, 45
steps in Project Plan section: major, 81–85, 87, 89–90, 92, 96; minor, 63, 64, 84–88, 90,–93, 96, 117
strengths (of a proposal), 10, 12, 98–101, 107–108, 113, 209
style: persuasive, xiii, 144, 156, 158, 162–165; plain, 144, 147, 150, 159, 167; subject, 145–147, 157

Style, Chapter 8 *passim*; 8, 144, 146, 158, 164, 178, 200, 208
style guides: using, 204
style sheets: for design, 178–180, 182
summaries, 201; executive, 200, 202

table of contents, 197, 202
timelines, 71, 91–92, 189
tone: in writing, 5, 10, 40, 89, 107, 115, 144, 158–159, 163–165, 167–168, 180, 198, 200, 208–209
topics, 33, 35, 45, 47, 53–54, 60, 151; importance of, 20; proposal, 33
travel: costs, 103, 131–132

values, xiii, 8, 39–40, 42, 45, 82–83, 85, 98, 112, 114–115, 121, 144, 206; as benefits, 113, 115, 118, 120
visual design, Chapter 11 *passim*; xiii; principles of, 4

www.ingramcontent.com/pod-product-compliance
Lightning Source LLC
Chambersburg PA
CBHW061130010526
44117CB00024B/3001